I0493823

High-Rise Database-Assisted Design 1.1 (HR_DAD_1.1): Concepts, Software, and Examples

Seymour M.J. Spence

Building and Fire Research Laboratory
National Institute of Standards and Technology
Gaithersburg, MD 20899-8611

February 2009

U.S. Department of Commerce
Dr. Gary Locke, *Secretary*

National Institute of Standards and Technology
Dr. Patrick D. Gallagher, *Deputy Director*

Disclaimer

Certain commercial entities, equipment, or materials may be identified in this document in order to describe an experimental procedure or concept adequately. Such identification is not intended to imply recommendation or endorsement by the National Institute of Standards and Technology, nor is it intended to imply that the entities, materials, or equipment are necessarily the best available for the purpose.

Abstract

This report documents the changes that have been made to the previous structure of the High-Rise Database-Assisted Design (HR_DAD) software, thereby developing the program HR_DAD_1.1 capable of calculating efficiently the response of tall buildings made up of many thousands of members. In particular the following upgrades on the previous software have been implemented: (1) Capability of performing time-domain calculations, requiring times of the order of hours, of the peak demand in each member of a real size structure (estimated time for the previous version of HR_ DAD on an AMD Athlon 64 processor 3000+ with 512MB was on the order of weeks); (2) Include as output of the software the peak demand of global response parameters such as inter-story drift and top floor acceleration for any number of column lines and locations on the top floor; (3) Allow time histories of floor displacements and top floor accelerations to be saved for selected wind directions and speeds; (4) Implement a general input format for the wind tunnel information concerning the Synchronous Multi-Pressure Sensing System measurements, therefore allowing for easy consideration of tall buildings with irregular geometries. (5) Predispose the software for the eventual inclusion of a multi-hazard approach to wind design of tall buildings, that is, consider various types of wind, including hurricanes, synoptic winds, and thunderstorms. The techniques, observations and theory that have been used to achieve these changes are documented together with the validation of the new software. The report includes a detailed user's manual.

Key words: Building technology; multi-hazard engineering; structural engineering; tall buildings; wind engineering.

Acknowledgments

This work was produced during the author's tenure as a Guest Researcher with the Structures Group, Building and Fire Research Laboratory (BFRL), on assignment from the Department of Civil and Environmental Engineering, University of Perugia, Italy. Dr. Massimiliano Gioffrè of the University of Perugia and CRIACIV (Italian Inter-University Research Center on Aerodynamics of Construction & Wind Engineering), and Dr. Emil Simiu of BFRL served as research advisers. The original HR_DAD software for which the enhancements described in this report were developed was created by Mihai Iancovici of the Bucharest Technical University of Civil Engineering and BFRL, William P. Fritz, BFRL, and René D. Gabbai, BFRL. Valuable contributions by Joseph A. Main, BFRL, are acknowledged with thanks. Mircea D. Grigoriu, Cornell University and BFRL, and Fahim Sadek, BFRL, provided helpful comments and suggestions on the format and content of the report. Permission to reproduce the cover photo was kindly given by Claudio Borri, Director, CRIACIV, University of Florence, Italy. The conceptual framework for the time-domain approach used in this work was developed by Emil Simiu, BFRL.

Contents

List of Figures

List of Tables

vii

1. INTRODUCTION

Modern tall buildings are increasingly characterized by irregular geometries. However, this newly found architectural freedom has complicated their structural design. Traditionally the design is carried out by considering a fairly large number of load combinations based on Equivalent Static Wind Loads (ESWLs) derived from High Frequency Force Balance (HFFB) wind tunnel testing and associated frequency domain analyses. This approach to the design of tall buildings against wind storms was developed primarily during the 1970s when the possibility of using Synchronous Multi-Pressure Sensing System (SMPSS) pressure measurements to characterize the external wind pressure field did not exist, and fast and efficient Ordinary Differential Equation (ODE) solvers were not available. This is not the case anymore as SMPSS measurements using up to 500 pressure taps are commonplace as are extremely fast and efficient ODE solvers.

The HFFB approach uses the fact that the expression for the base moments in sway coincides with the respective generalized forces, provided that the fundamental sway modes depend linearly on height [1, 2]. That approach has the following shortcomings. First, it measures only the base moments, and therefore provides no information on the effective distribution of the wind load over the building's height. This is an obstacle to calculating the mean and background responses [3, 4], as well as to implementing the various mode correction schemes required if the fundamental sway modes do not vary linearly with height or are coupled, as is the case if the center of mass and the elastic center do not coincide [5, 6, 7, 8, 9, 10, 11, 12, 13, 14]. Therefore, in the case of HFFB measurements, the only path is to guess the load distribution, which introduces approximations that can be particularly crude in the presence of aerodynamic interference effects due to neighboring tall buildings. Second, HFFB does not provide the information needed to account for the effect of higher modes of vibration when calculating the mean, background, and resonant response as only the first three generalized forces may be estimated. Third, corrections are always required for the fundamental torsional mode, which would have to vary uniformly with height if the fundamental generalized moment in torsion were to coincide with the base torsional moment. These corrections depend on assumptions concerning the wind load distribution on the exterior surface of the building. Fourth, because the spectral approach commonly used with the HFFB approach entails total loss of phase information, difficulties arise when attempting to account for joint effects of responses in the fundamental modes. These difficulties are addressed by considering multiple load combinations that have little or no theoretical foundation and therefore introduce additional approximations, as well as massively increasing the designer's computational tasks. Fifth, inherent to all frequency domain methods, is the inability to account for non-Gaussian features of the response, which can be significant [4,15].

All these shortcomings are overcome by adopting a time-domain approach with associated Synchronous Multi-Pressure Sensing System (SMPSS) pressure measurements [16, 17, 18]. SMPSS pressure measurements allow the distribution of wind loads over the building to be estimated. Therefore the mean and background responses can be easily and transparently estimated, automatically accounting for spatial pressure correlations and aerodynamic interference effects. Also, because the loads over the building height are known, any number of modes may be considered in the resonant response. This technique is also independent of whether the building exhibits uncoupled and linear mode shapes or the coupled modes with nonlinear shapes, and can be applied to structures with simple or complex geometries. The time

1

domain automatically accounts for the correlation between the modal responses, for phase relationships among various wind effects, and for any non-Gaussian effects.

Time-domain techniques are now possible owing to the availability of fast ordinary differential equation solvers. Nevertheless, they can be time-consuming, especially when considering the numerous constraints that must be satisfied in the design of tall buildings. It is therefore paramount for the success of any time-domain based analysis package to be very efficient. This will allow the designer to tap into the advantages that are inherent in this approach.

This report documents the changes that have been made to the previous structure of the HR_DAD software (see www.nist.gov/wind, II A, Wind Effects on Flexible Buildings) with the aim of making this a program capable of calculating efficiently the response of large scale real world tall buildings made up of many thousands of members. In particular the following upgrades on the previous software have been implemented:

- Capability of performing calculations, requiring times of the order of hours, of the peak demand in each member of a real size structure (estimated time for the previous version of HR_ DAD on an AMD Athlon 64 processor 3000+ with 512MB was on the order of weeks).

- Include as output of the software the peak demand of global response parameters such as inter-story drift and top floor acceleration for any number of column lines and locations on the top floor.

- Allow time histories of floor displacements and top floor accelerations to be saved for selected wind directions and speeds.

- Implement a general input format for the wind tunnel information concerning the SMPSS measurements, therefore allowing for easy consideration of tall buildings with irregular geometries.

- Predispose the software for the eventual inclusion of a multi-hazard approach to wind design of tall buildings, that is, consider various types of wind, including hurricanes, synoptic winds, and thunderstorms.

In the following sections the techniques, observations and theory that have been used to achieve these changes will be documented together with the verification of the updated software, which can be found on Appendix A. The modified software is named HR_DAD_1.1, and is incorporated in the site www.nist.gov/wind in Section II A under the heading "updated HR_DAD software (HR_DAD_1.1)."

2. GENERAL STRUCTURE OF HR_DAD_1.1

The general structure of the software has remained similar to that of the previous version, HR_DAD, documented in www.nist.gov/wind, Section II A, Wind Effects on Flexible Buildings [16]. The HR_DAD_1.1 software consists of 34 MATLAB files arranged as shown in Fig. 2.1.

The program is started by typing "HR_DAD" at the MATLAB command prompt. This file initializes all the variables and opens up 10 pages: 2 introductory pages (Page_Main and Page_Open) and 8 other pages (Page_One through Page_Eight) that run the program. Each page contains the eight menu tabs located along the left-hand side of the graphical user interface. Selection of one of these tabs brings that page to the top of all the opened pages. Pages 1 to 5 contain areas to input values for the variables and open files pertinent to the program. Purple icons within each page that contain the symbol "?" provide additional information for the specific item. For example, the menu labelled "1.Building Info.", which calls the Page_One.m routine, contains an input box for the variable "RNmodes". Clicking on the purple information icon produces a pop-up window that indicates that the variable name within the MATLAB program is RNmodes, that its matrix size is 1 x 1 (i.e., it is a constant), and that the variable

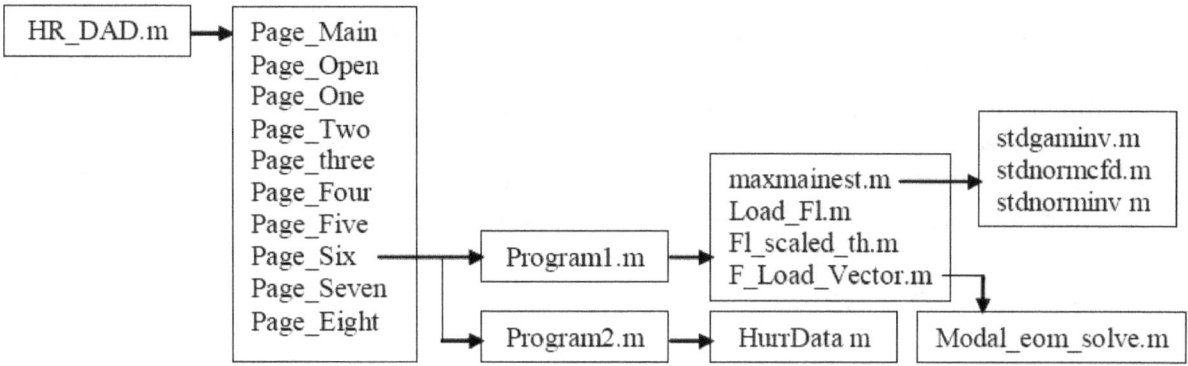

Figure 2.1. HR_DAD_1.1 program structure

specifies the number of vibration modes considered in the analysis. The user defines the necessary variables in pages one through five, which automatically change the values of the (global) variables within the program. The pages and the variables can be selected in any order. The default values specified in the HR_DAD_1.1.m file are otherwise used.

The actual calculations are performed in page six by selecting the "Run1" and "Run2" buttons, which run the files called Program1.m and Program2.m, respectively (refer to Fig. 2.1). Program1.m uses the relative variables to generate the *response surfaces*, defined in this report as surfaces whose ordinates are peak responses and whose abscissas are wind speeds and wind directions, for specified members, inter-story drift (for any number of column lines input on page five), and top floor accelerations (for any number of locations input on page five). The methods used to calculate the peak response are differentiated depending on whether a member response or global response is being considered. In view of the potentially enormous number of members composing tall buildings (in the order of many thousands) the local member level responses are calculated by assuming them to be Gaussian. This allows the peak response to be calculated with great efficiency directly from the covariance matrix of the forces acting on the centers of mass of each floor. The calculation technique is fully explained in section 5.4. The global response on the other hand consists of a considerably smaller number (on the order of hundreds) of demands of interest, therefore allowing the peaks of the relevant times series to be estimated either as observed peaks or by the modified statistical Rice method, see page five.

For each wind direction WD and wind speed WS (defined in page three) several functions are used, some of which call other functions as shown in Fig. 2.1. Comments within the MATLAB script files describe details for these functions. The routine in Program1.m establishes three 3D arrays of peak values that have dimensions WS × WD × (number of members or column lines or locations where top floor acceleration is calculated) and saves them under the file locations and names specified in page six. Each face of an array is a discrete representation of the response surfaces.

The 3D arrays saved in Program1.m are used in Program2.m. The routine in Program2.m cycles through each member, inter-story drift and acceleration constraint, loads the saved response surfaces from Run1 (Program1.m), loads the particular hurricane wind speed database (specified in page five), and calculates the peak wind effect for each MRI (specified in page three and distinguished between those pertaining to members and those to be used for the global responses). The peak wind demands as a function of the MRIs are then saved for each member, inter-story drift set and top floor acceleration point under the file location and name specified in page six.

Page seven summarizes key input variables while page eight, "8. Save/Load Data", allows the user to save the current set of variables to a specified file location or load a previously saved file. Finally, the "Exit" button closes all pages and exits the HR_DAD_1.1 program. The variables, however, still remain in memory. The following Sections will outline the theory behind the various parts of HR_DAD_1.1. Reference will be made to the relevant input quantities used during the implementation of the routines and to the eventual output.

3. ANALYSIS FRAMEWORK

3.1 Wind tunnel input (page 2.m)

The wind loads and related information are input on page three (Fig. 2.1). The loads consist of the time histories of the model scale floor loads acting at the center of mass of each of the N floors. The loads are arranged in a $3N \times 1$ vector $\{F(t)\} = \begin{bmatrix} \mathbf{F}_x^T(t) & \mathbf{F}_y^T(t) & \mathbf{F}_\theta^T(t) \end{bmatrix}^T$ where the loads in the directions x, y and θ are given by the sub-vectors $\mathbf{F}_x(t) = \left(F_{1x}(t), F_{2x}(t), .. F_{Nx}(t) \right)^T$, $\mathbf{F}_x(t) = \left(F_{1y}(t), F_{2y}(t), .. F_{Ny}(t) \right)^T$ and $\mathbf{F}_x(t) = \left(F_{1\theta}(t), F_{2\theta}(t), .. F_{N\theta}(t) \right)^T$. These loads are estimated through wind tunnel tests on rigid scale models equipped with a number of simultaneously measured pressure taps that allow for the characterization of the external wind pressure field through non-dimensional pressure coefficients. It is assumed throughout that aeroelastic effects are negligible. The transformation of the external pressure field into concentrated loads acting at the centers of mass may be performed following procedures deemed adequate by the user. The resulting model scale floor loads must be scaled before they can be applied to the dynamic system. This is done automatically in HR_DAD_1.1 by respecting the similitude on the reduced frequency:

$$\frac{f_m D_m}{V_m} = \frac{f_p D_p}{V_p} \tag{1}$$

where D denotes a characteristic dimension of the structure, f denotes the sampling frequency, V denotes the mean wind velocity at a consistent height (e.g., roof height), and the subscripts m and p denote "model" and "prototype," respectively. Letting $\lambda_L = D_m/D_p$ denote the length scale of the wind tunnel model, the prototype sampling frequency can be expressed as follows by rearranging Eq. (1):

$$f_p = f_m \left(\frac{V_p}{V_m} \right) \lambda_L \tag{2}$$

The magnitude of the loads must also be scaled. This is simply done by multiplying the time series of the model loads by $\left(V_p / V_m \right)^2$. All scaling is performed during the execution of program1.m. The information necessary for this is input in page three.

3.2 Mechanical model (program1.m and modal_eom_solve.m)

The global behavior of tall buildings can be modeled by an equivalent dynamical system considering each floor to have three degrees of freedom (i.e. *x*- and *y*-displacement, relative to the ground, of the center of mass, and θ-rotation about a vertical axis through the center of mass). Under these hypotheses the dynamic equilibrium of an *N*-story building with mass and stiffness eccentricities which can vary from floor to floor is given by:

$$[M]\{\ddot{U}(t)\} + [C]\{\dot{U}(t)\} + [K]\{U(t)\} = \{F(t)\} \tag{3}$$

In which:

$$[M] = \begin{bmatrix} \mathbf{M} & \mathbf{0} & \mathbf{0} \\ \mathbf{0} & \mathbf{M} & \mathbf{0} \\ \mathbf{0} & \mathbf{0} & \mathbf{J} \end{bmatrix} \tag{4a}$$

$$[K] = \begin{bmatrix} \mathbf{K_{xx}} & {}_\theta\mathbf{K_{xy}} & \mathbf{K_x} \\ \mathbf{K_{xy}^T} & {}_\theta\mathbf{K_{yy}} & \mathbf{K_y} \\ \mathbf{K_{x\theta}^T} & \mathbf{K_{y\theta}^T} & \mathbf{K_{\theta\theta}} \end{bmatrix} \tag{4b}$$

$$[C] = \begin{bmatrix} \mathbf{C_{xx}} & {}_\theta\mathbf{C_{xy}} & \mathbf{C_x} \\ \mathbf{C_{xy}^T} & {}_\theta\mathbf{C_{yy}} & \mathbf{C_y} \\ \mathbf{C_{x\theta}^T} & \mathbf{C_{y\theta}^T} & \mathbf{C_{\theta\theta}} \end{bmatrix} \tag{4c}$$

$$\{U(t)\} = \begin{bmatrix} \mathbf{X}^T(t) & \mathbf{Y}^T(t) & {}^T(t) \end{bmatrix}^T \tag{4d}$$

where the displacement response sub-vectors are $\mathbf{X(t)}=(x_1(t),x_2(t),...,x_N(t))^T$, $\mathbf{Y(t)}=(y_1(t),y_2(t),...,y_N(t))^T$ and $\mathbf{\Theta(t)}=(\theta_1(t),\theta_2(t),...,\theta_N(t))^T$; the mass sub-matrix is $\mathbf{M}=diag[m_i]$ for $i=1,...,N$, in which m_i is the lumped mass of floor i, while the sub-matrix of mass moment of inertias of the floor diaphragms is $\mathbf{J}=diag[J_i]$ for $i=1,...,N$, in which J_i represents the polar moment of inertia of floor i about a vertical axis through the center of mass. $\mathbf{K_{xx,}}$ $\mathbf{K_{yy,}}$ $\mathbf{K_{\theta\theta,}}$ $\mathbf{K_{yx,}}$

5

$\mathbf{K}_{x\theta}$, $\mathbf{K}_{y\theta}$ and $\mathbf{C}_{xx}$, $\mathbf{C}_{yy}$, $\mathbf{C}_{\theta\theta}$, $\mathbf{C}_{xy}$, $\mathbf{C}_{x\theta}$, $\mathbf{C}_{y\theta}$ are the sub-matrixes of the stiffness and damping of the building respectively.

These generally coupled equations of motion can be solved through modal analysis. In this framework the modal frequencies ω_k and mode shapes $\{\Phi_k\}$ for $k=1,2...3N$ are the roots and non-trivial solution of the following equations respectively:

$$\det(-\omega^2[M]+[K]) = 0 \text{, or} \tag{5a}$$

$$(-\omega_k^2[M]+[K])\{\Phi_k\} = 0 \tag{5b}$$

From the structure of Eq. 3 it is obvious that if the centers of mass and stiffness for each floor are coincident and lie on a vertical axis then the system will be uncoupled and the building will experience uncoupled vibration modes in three principal directions, two orthogonal translational directions and a rotational direction [1,2]. Otherwise the building will, in general, experience 3D coupled modes. Due to the orthogonality of the mode shapes, if damping is considered then the generally coupled system of Eq. 2 maybe transformed into a set of $j=1,2,....,3N$ uncoupled equations in the generalized coordinates q_j:

$$\ddot{q}_j(t)+2\xi_j\omega_j\dot{q}_j(t)+\omega_j^2 q_j(t) = \frac{Q_j(t)}{\{\Phi_j\}^T[M]\{\Phi_j\}} \tag{6}$$

where ξ_j, and ω_j are the damping ratio and circular frequency of mode j, while $Q_j(t)$ is the generalized force given by:

$$Q_j(t) = \{\Phi_j\}^T \{F(t)\} \tag{7}$$

Therefore from the knowledge of the global mass matrix, circular frequencies, damping ratios, mode shapes and scaled time varying floor loads, the modal equations of motion can be solved. It is this information input on pages 1, 2 and 4 (Fig. 2.1) that is used to define the modal equations of motion (Eq. 6) that are directly integrated by the MATLAB function modal_eom_solve.m during the execution of program1.m. The number of modes to be considered in the solution is user defined. Because the distribution of the wind loads with height is known, any number of modes may be included in the analysis.

Following this framework the response displacements and accelerations in the global reference system can be expressed in terms of the contributions of $j=1,2,...,n\leq3N$ modes. The displacements for example are given by:

$$\{U(t)\} = [\Phi]\{q(t)\} \tag{8}$$

where $[\Phi]$ is the matrix consisting of the n vectors $\{\Phi_j\}$ for $j=1,2,...,n\leq3N$ while $\{q\} = (q_1(t),q_2(t),...,q_n(t))^T$ denotes the vector of generalized displacements. Likewise the accelerations are given by:

$$\{\ddot{U}(t)\} = [\Phi]\{\ddot{q}(t)\}$$

(9)

4. GLOBAL RESPONSE SURFACES: INTER-STORY DRIFT AND TOP FLOOR ACCELERATION (PROGRAM1.M)

From the global response, Eq. 6, it is possible to construct the peak response surfaces for the inter-story drift and top floor acceleration. For example, given a certain incident wind direction WD and mean velocity and the top of the building WS, the peak inter-story drift for the k_{th} set of points and j_{th} floor and in direction $s=x,y$ is given by:

$$d_{jks}^{(WS,WD)}(t) = \frac{[x_j^{(WS,WD)}(t) + \Delta_{kjs}\theta_j^{(WS,WD)}(t)] - [x_{j-1}^{(WS,WD)}(t) + \Delta_{kj-1s}\theta_{j-1}^{(WS,WD)}(t)]}{h_j}$$

(10)

in which $\Delta_{kjs} = -D_{kjy}$ for $s=x$ while for $s=y$ $\Delta_{kjs} = D_{kjx}$ for $j=1,2,...,N$, where D_{kjx} and D_{kjy} are shown in Fig. 4.2, kj and $kj-1$ are the two points belonging to the set k for which the inter-story drift is to be controlled (Fig. 2.1) and h_j is the height of the jth story.

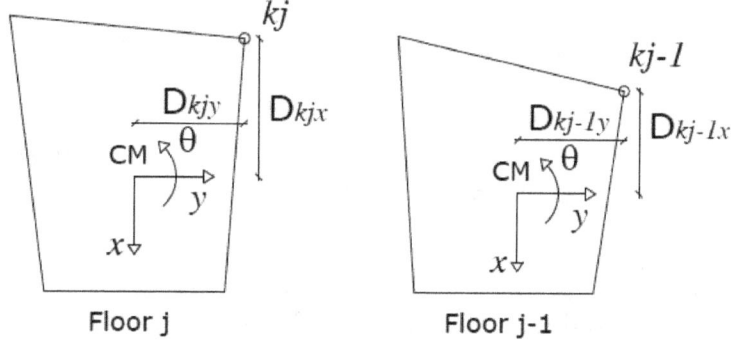

Figure 4.1. Position parameters D_{kjx} and D_{kjy} for floors j and j-1 and points kj and kj-1.

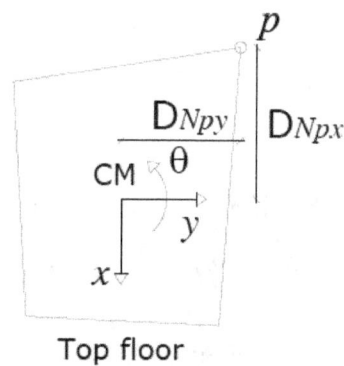

Figure 4.2. Position parameters D_{Npx} and D_{Npy} for the top floor.

The top floor acceleration is estimated in a similar fashion. Given a point $p=1,2,...,P$ belonging to the top floor the acceleration is given by:

$$a_{Ns}^{(WS,WD)}(t) = \ddot{x}_N^{(WS,WD)}(t) + \Delta_{Ns}\ddot{\theta}_N^{(WS,WD)}(t) \tag{11}$$

in which $\Delta_{Np} = -D_{pNy}$ for $s=x$ while for $s=y$ $\Delta_{Ns} = D_{pNx}$ for $p=1,2,...,P$, where D_{pNy} and D_{pNx} are shown in Fig. 4.2.

From the time series of the displacements and accelerations it is possible to evaluate the peak response in a number of ways. In particular HR_DAD_1.1 software allows the selection (page five) between considering the observed peak or fitting a marginal distribution to the record (subroutine maxminest.m). The estimation of the response surface is achieved by solving the equations of motion for a number of wind directions (WD) and wind speeds (WS). Each surface is saved in a 3D array, each face of which represents a response surface for a particular component. The quantities D_{kjx}, D_{kjy} and h_j for $j=1,2,...,N$ are input via a 3 x (number of sets x N) matrix on page five (Fig. 2.1) as are D_{pNy} and D_{pNx} via a 2 x (number of points) matrix.

5. LOCAL RESPONSE SURFACES: FORMULATION OF THE STRENGTH REQUIREMENTS

5.1 Influence coefficients

The time varying internal forces induced in any member by a given incident wind direction $\gamma =$ WD and mean velocity at the top of the building $\overline{V} =$ WS may be obtained from the following relationships, which for the sake of brevity are shown only for the axial forces and bending moments:

$$N_i^{\overline{V}_\gamma}(t) = \sum_{j=1}^{N} n_{jxi}[F_{jx}^{\overline{V}_\gamma}(t) - m_j\ddot{x}_j^{\overline{V}_\gamma}(t)] + n_{jyi}[F_{jy}^{\overline{V}_\gamma}(t) - m_j\ddot{y}_j^{\overline{V}_\gamma}(t)] + n_{j\theta i}[F_{j\theta}^{\overline{V}_\gamma}(t) - I_j\ddot{\theta}_j^{\overline{V}_\gamma}(t)] \tag{12a}$$

$$M_{iX}^{\overline{V}_\gamma}(t) = \sum_{j=1}^{N} m_{jXxi}[F_{jx}^{\overline{V}_\gamma}(t) - m_j\ddot{x}_j^{\overline{V}_\gamma}(t)] + m_{jXyi}[F_{jy}^{\overline{V}_\gamma}(t) - m_j\ddot{y}_j^{\overline{V}_\gamma}(t)] + m_{jX\theta i}[F_{j\theta}^{\overline{V}_\gamma}(t) - I_j\ddot{\theta}_j^{\overline{V}_\gamma}(t)] \tag{12b}$$

$$M_{iY}^{\overline{V}_\gamma}(t) = \sum_{j=1}^{N} m_{jYxi}[F_{jx}^{\overline{V}_\gamma}(t) - m_j\ddot{x}_j^{\overline{V}_\gamma}(t)] + m_{jYyi}[F_{jy}^{\overline{V}_\gamma}(t) - m_j\ddot{y}_j^{\overline{V}_\gamma}(t)] + m_{jY\theta i}[F_{j\theta}^{\overline{V}_\gamma}(t) - I_j\ddot{\theta}_j^{\overline{V}_\gamma}(t)] \tag{12c}$$

in which $F_{js}^{\overline{V}_q}(t)$ for $s=x,y,\theta$ are the components of j_{th} floor load due to the external wind loads acting at the center of mass while $m_j\ddot{x}_j(t)$, $m_j\ddot{y}_j(t)$ and $I_j\ddot{\theta}_j(t)$ are the inertial force components.

The quantities m_{jXxi}, m_{jYxi}, m_{jXyi}, m_{jYyi} $m_{jX\theta i}$, $m_{jY\theta i}$, n_{jqi}, n_{jxi} and n_{jyi} are the influence coefficients necessary for the calculation of the internal forces. In particular m_{jXxi} and m_{jYxi} are the bending moments M_{jX} and M_{jY} about the local axis X and Y of member i (Fig. 5.1) due to a unit force acting through the center of mass in the x-direction at floor j, while m_{jXyi} and m_{jYyi} are similar moments due to a unit force acting through the center of mass in the y-direction at floor j;

$m_{jX\theta i}$ and $m_{jY\theta i}$ are the moments due to a unit torque about the mass center at floor j while n_{jqi} is the axial force due to this unit torque; n_{jxi} and n_{jyi} are the axial forces due to a unit load in the directions x and y applied to the center of mass of floor j. Obviously the influence coefficients depend on the section of the member in which they are calculated. HR_DAD_1.1 considers three sections for each member in which to calculate the internal forces and thereafter verify the demand-to-capacity index (i.e., the left-hand side of the interaction equation, see Section 5.2) of the member. In particular the initial, final and mid sections are chosen, that is, sections 1, 2 and 3 of Fig. 5.1.

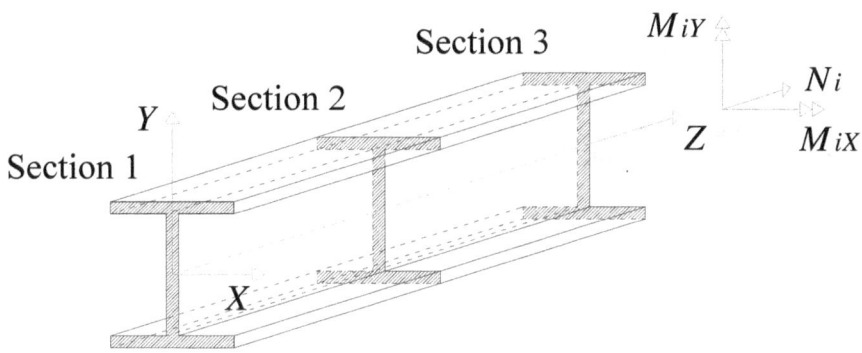

Figure 5.1. Member i.

HR_DAD_1.1 does not calculate the influence coefficients. They must be obtained separately using any commercially available structural analysis program such as SAP 2000. This does not present any particular difficulty. By definition the influence coefficients are simply the internal forces in three sections of each member due to a statically applied unit load. The only possible difficulty could lie in the number of load cases, $3N$ where N is the number of floors, which must be solved. However this process may be rendered extremely efficient by writing a subroutine that interacts with the structural analysis program and automatically prepares the input file for HR_DAD_1.1. It should be observed that even if this not done, due to the extreme efficacy of modern structural analysis software the calculation of the influence coefficients will only take a couple of hours or so. Note that the number of influence coefficients is quite considerable when storage issues are being considered. To avoid problems the coefficients are arranged in a 3D array, each face of which contains the influence coefficients for the three sections associated with a particular member. Information on the exact structure of this array maybe found by pressing on the "?" icon on page three of the HR_DAD_1.1 Graphical User Interface (GUI) under the heading dynamic influence coefficients.

5.2 Interaction formulas

In the design of tall buildings, however, it is not individual internal forces that are of interest but rather combinations of internal forces governed by interaction equations and load combinations. In the case of steel structures for example, the interaction formulas given by AISC (2001) for the design of member i are:

$$\frac{N_{if}^{\bar{V}_{\gamma}}(t)}{\phi N_{ni}} \geq 0.2 \Rightarrow b_{if}^{\bar{V}_{\gamma}}(t) = \frac{N_{if}^{\bar{V}_{\gamma}}(t)}{\phi N_{ni}} + \frac{8}{9}\left(\frac{M_{iXf}^{\bar{V}_{\gamma}}(t)}{\phi_b M_{niX}} + \frac{M_{iYf}^{\bar{V}_{\gamma}}(t)}{\phi_b M_{niY}}\right) \leq 1 \tag{13a}$$

$$\frac{N_{if}^{\bar{V}_{\gamma}}(t)}{\phi N_{ni}} < 0.2 \Rightarrow b_{if}^{\bar{V}_{\gamma}}(t) = \frac{N_{if}^{\bar{V}_{\gamma}}(t)}{2\phi N_{ni}} + \frac{M_{iXf}^{\bar{V}_{\gamma}}(t)}{\phi_b M_{niX}} + \frac{M_{iYf}^{\bar{V}_{\gamma}}(t)}{\phi_b M_{niY}} \leq 1 \tag{13b}$$

where N_{ni}, M_{niX} and M_{niY} are the nominal axial and flexural strengths of member i, ϕ and ϕ_b are axial and flexural resistance factors while $N_{if}^{\bar{V}_{\gamma}}(t)$, $M_{iXf}^{\bar{V}_{\gamma}}(t)$ and $M_{iYf}^{\bar{V}_{\gamma}}(t)$ are the total internal forces due to specified factored combinations (hence the subscript f), with dynamic wind load contributions calculated from Eqs. 12a-c. Examples of appropriate load combinations can be found in the ASCE 7-05 Standard:

$$1.2D + 1.6L \tag{14a}$$

$$1.2D + 1.0L + 1.6W \tag{14b}$$

$$0.9D + 1.6W \tag{14c}$$

in which D is the dead load, L is the live load while W is the wind load.

Obviously equations 13a-b represent two distinct time series. Whether 13a or 13b should be considered depends on the value of $N_{if}^{\bar{V}_{\gamma}}(t)/\phi N_{ni}$. To construct the response surface it is necessary to calculate the peak of $b_{if}^{\bar{V}_{\gamma}}(t)$ for a sufficient number of wind speeds and directions. In the following two paragraphs methods for calculating these peaks are presented.

5.3 Formulation based on time history analysis

To better understand the structure of $b_{if}^{\bar{V}_{\gamma}}(t)$ it is convenient to rewrite Eqs. 13a-b, dividing them into a mean static response and a zero mean fluctuating response. For the sake of brevity only Eq. 13b with combination 14b is shown. The other cases are equivalent.

$$\frac{N_{if}^{\bar{V}_{\gamma}}(t)}{\phi N_{ni}} < 0.2 \Rightarrow b_{if}^{\bar{V}_{\gamma}}(t) = \left[\frac{1.6\bar{N}_i^{\bar{V}_{\gamma}} + N_{if}^s}{2\phi N_{ni}} + \frac{1.6\bar{M}_{iX}^{\bar{V}_{\gamma}} + M_{iXf}^s}{\phi_b M_{niX}} + \frac{1.6\bar{M}_{iY}^{\bar{V}_{\gamma}} + M_{iYf}^s}{\phi_b M_{niY}}\right] +$$

$$+ \left[\frac{1.6}{2\phi N_{ni}}N_i^{\bar{V}_{\gamma}}(t) + \frac{1.6}{\phi_b M_{niX}}M_{iX}^{\bar{V}_{\gamma}}(t) + \frac{1.6}{\phi_b M_{niY}}M_{iX}^{\bar{V}_{\gamma}}(t)\right] \leq 1 \tag{15}$$

in which N_{if}^s, M_{iXf}^s and M_{iYf}^s are the factored internal forces; $\bar{N}_i^{\bar{V}_{\gamma}}$, $\bar{M}_{iX}^{\bar{V}_{\gamma}}$ and $\bar{M}_{iY}^{\bar{V}_{\gamma}}$ are the mean internal forces due to the external wind loading for $\bar{V}_{\gamma}(H)$ while $N_i^{\bar{V}_{\gamma}}(t)$, $M_{iX}^{\bar{V}_{\gamma}}(t)$ and $M_{iX}^{\bar{V}_{\gamma}}(t)$ are the corresponding zero mean fluctuating components.

10

From the structure of Eq. 15 it is obvious that in order to evaluate $b_{if}^{\bar{V}_\gamma}(t)$ for any number of combinations and interaction formulas it is necessary to calculate the time histories of the fluctuating internal forces just once for each section of interest for member i. Once the time history of $b_{if}^{\bar{V}_\gamma}(t)$ is available the peak may be calculated. In particular, for non-Gaussian time histories, the method included in the site www.nist.gov/wind, Sect. III B may be used. However, due to the large number of elements belonging to the structure of a typical tall building, and given the need to evaluate the response for a sufficient number of wind speeds and directions, the calculation of the time histories associated with the fluctuating wind load causes a combinatorial explosion. To avoid this, data compression techniques can be used. The evaluation of the time histories can be completely avoided if the response can be modeled as Gaussian – which is the case for effects associated with global contributions to the response, -- leading to a significant computational advantage. This approach is the method adopted in HR_DAD_1.1 and is fully explained in the following section.

5.4 Formulation based on the expected peak (program1.m)

If $b_{if}^{\bar{V}_\gamma}(t)$ is assumed to be Gaussian, its expected peak is given by the well known Rice formula (Davenport, 1964) as:

$$[b_{if}^{\bar{V}_\gamma}(t)]_{peak} = \bar{b}_{if}^{\bar{V}_\gamma} + g\sigma_{b_{if}^{\bar{V}_\gamma}} \tag{16}$$

in which $\bar{b}_{if}^{\bar{V}_\gamma}$ is the mean response; g is the peak factor while $\sigma_{b_{if}^{\bar{V}_\gamma}}$ is the standard deviation of $b_{if}^{\bar{V}_\gamma}(t)$. By the definition of the variance, it follows from Eq. 15 that the variance of $b_{if}^{\bar{V}_\gamma}(t)$ may be written as:

$$\sigma_{b_{if}^{\bar{V}_\gamma}}^2 = E\left[\left(k_1 N_i^{\bar{V}_\gamma}(t) + k_2 M_{iX}^{\bar{V}_\gamma}(t) + k_3 M_{iX}^{\bar{V}_\gamma}(t)\right)^2\right] \tag{17}$$

where:

$$k_1 = \frac{1.6}{2\phi N_{ni}} \tag{18a}$$

$$k_2 = \frac{1.6}{\phi_b M_{niX}} \tag{18b}$$

$$k_3 = \frac{1.6}{\phi_b M_{niY}} \tag{18c}$$

Therefore, in view of the linearity of the expectation operator E, it is possible to the express the standard deviation as:

11

$$\sigma_{b_{if}^{\bar{V}_\gamma}} = \sqrt{k_1^2 C_{N_i^{\bar{V}_\gamma} N_i^{\bar{V}_\gamma}} + k_2^2 C_{M_{iX}^{\bar{V}_\gamma} M_{iX}^{\bar{V}_\gamma}} + k_3^2 C_{M_{iY}^{\bar{V}_\gamma} M_{iY}^{\bar{V}_\gamma}} + 2k_1 k_2 C_{N_i^{\bar{V}_\gamma} M_{iX}^{\bar{V}_\gamma}} + 2k_1 k_3 C_{N_i^{\bar{V}_\gamma} M_{iY}^{\bar{V}_\gamma}} + 2k_2 k_3 C_{M_{iX}^{\bar{V}_\gamma} M_{iY}^{\bar{V}_\gamma}}} \qquad (19)$$

where $C_{N_i^{\bar{V}_\gamma} N_i^{\bar{V}_\gamma}}$, $C_{M_{iX}^{\bar{V}_\gamma} M_{iX}^{\bar{V}_\gamma}}$, $C_{M_{iY}^{\bar{V}_\gamma} M_{iY}^{\bar{V}_\gamma}}$, $C_{N_i^{\bar{V}_\gamma} M_{iX}^{\bar{V}_\gamma}}$, $C_{N_i^{\bar{V}_\gamma} M_{iY}^{\bar{V}_\gamma}}$ and $C_{M_{iX}^{\bar{V}_\gamma} M_{iY}^{\bar{V}_\gamma}}$ are the various covariance coefficients between the fluctuating components of the internal forces. These coefficients can be calculated directly from the covariance matrix of the global forces $\mathbf{F_x}$ acting at the centers of mass of each floor and the influence coefficients of the internal forces. For example, the covariance coefficient $C_{N_i^{\bar{V}_\gamma} M_i^{\bar{V}_\gamma}}$ between the time varying axial force and bending moment around the local X axis in member i may be written as:

$$C_{N_i^{\bar{V}_\gamma} M_i^{\bar{V}_\gamma}} = E\left[N_i^{\bar{V}_\gamma}(t) M_{iX}^{\bar{V}_\gamma}(t) \right] \qquad (20)$$

By substituting Eqs. 12a-b into Eq. 20, the right-hand side of Eq. 20 may be rewritten in the following form:

$$E\left[\left(\sum_{j=1}^{N} n_{jxl}[F_{jx}^{\bar{V}_\gamma}(t) - m_j \ddot{x}_j^{\bar{V}_\gamma}(t)] + n_{jyl}[F_{jy}^{\bar{V}_\gamma}(t) - m_j \ddot{y}_j^{\bar{V}_\gamma}(t)] + n_{j\theta l}[F_{j\theta}^{\bar{V}_\gamma}(t) - I_j \ddot{\theta}_j^{\bar{V}_\gamma}(t)] \right) \right.$$

$$\left. \left(\sum_{j=1}^{N} m_{jXxi}[F_{jx}^{\bar{V}_\gamma}(t) - m_j \ddot{x}_j^{\bar{V}_\gamma}(t)] + m_{jXyi}[F_{jy}^{\bar{V}_\gamma}(t) - m_j \ddot{y}_j^{\bar{V}_\gamma}(t)] + m_{jX\theta i}[F_{j\theta}^{\bar{V}_\gamma}(t) - I_j \ddot{\theta}_j^{\bar{V}_\gamma}(t)] \right) \right] \qquad (21)$$

which may in turn be written as:

$$\sum_{j=1}^{N} \left(\sum_{k=1}^{N} n_{jxi} m_{kXxi} E\left[\left(F_{jx}^{\bar{V}_\gamma}(t) - m_j \ddot{x}_j^{\bar{V}_\gamma}(t) \right) \left(F_{kx}^{\bar{V}_\gamma}(t) - m_k \ddot{x}_k^{\bar{V}_\gamma}(t) \right) \right] \right) +$$

$$\sum_{j=1}^{N} \left(\sum_{k=1}^{N} n_{jxi} m_{kXyi} E\left[\left(F_{jx}^{\bar{V}_\gamma}(t) - m_j \ddot{x}_j^{\bar{V}_\gamma}(t) \right) \left(F_{ky}^{\bar{V}_\gamma}(t) - m_k \ddot{y}_k^{\bar{V}_\gamma}(t) \right) \right] \right) + \ldots \qquad (22)$$

where, for example, $E\left[\left(F_{jx}^{\bar{V}_\gamma}(t) - m_j \ddot{x}_j^{\bar{V}_\gamma}(t) \right) \left(F_{kx}^{\bar{V}_\gamma}(t) - m_k \ddot{x}_k^{\bar{V}_\gamma}(t) \right) \right]$ is simply the covariance coefficient between the total forces acting on the center of mass of floors j and k in direction x. Therefore there is no need to calculate the individual time histories of the internal forces, thus avoiding a combinatorial explosion.

As previously observed, whether Eq. 13a or 13b should be considered depends on the value assumed by $N_{if}^{\bar{V}_\gamma}(t)/\phi N_{ni}$ at the instant in which the value of $b_{if}^{\bar{V}_\gamma}(t)$ is desired. However under the formulation presented in this paragraph time is eliminated therefore it is not possible to know the value the of $N_{if}^{\bar{V}_\gamma}(t)/\phi N_{ni}$ that accompanies the expected peak of $b_{if}^{\bar{V}_\gamma}(t)$. In HR_DAD_1.1 this problem is solved by considering the expected maximum and minimum of $N_{if}^{\bar{V}_\gamma}(t)/\phi N_{ni}$ that accompany $b_{if}^{\bar{V}_\gamma}(t)$. More precisely if $\left[\left|N_{if}^{\bar{V}_\gamma}(t)\big/\phi N_{ni}\right|\right]_{Expected\ max} < 0.2$, then only the expected peak

of 13b is considered, while if $\left[N_{if}^{\bar{V}_\gamma}(t)\big/\phi N_{ni}\right]_{Expected\ max}$ and $\left[N_{if}^{\bar{V}_\gamma}(t)\big/\phi N_{ni}\right]_{Expected\ min} \geq 0.2$ or

$\left[N_{if}^{\bar{V}_\gamma}(t)\big/\phi N_{ni}\right]_{Expected\ max}$ and $\left[N_{if}^{\bar{V}_\gamma}(t)\big/\phi N_{ni}\right]_{Expected\ min} \leq 0.2$, then only 13a is considered. In all

other cases both the expected peaks of 13a and 13b are calculated, with the larger of the two being saved as representative of the peak demand.

To implement this formulation the influence coefficients must be input. The number of coefficients depends on the number of sections for which the peak response is to be calculated along the member. At present in HR_DAD_1.1 a maximum of three sections, the two extreme sections and the center section, are considered for every member. Therefore for each member a total of 6 (number if internal forces) × 3 (number of sections to be considered) × N (number of floors) have to be considered. To input all these coefficients a 3D array is used, in which each face contains the necessary influence coefficients for the three sections of a particular section. This array is input on page three (Fig. 2.1). During the execution of program1.m a subroutine will run through the list of members, also input on page three, and calculate the various covariance coefficients and the expected peak. To improve speed a distinction is made between tension and compression elements, and beams and columns, as the number of sections for which

it is necessary to calculate $[b_{if}^{\bar{V}_\gamma}(t)]_{peak}$ depends of the type of element under consideration. For

each wind direction γ = WD and wind speed $\bar{V}$ = WS only the maximum $[b_{if}^{\bar{V}_\gamma}(t)]_{peak}$ occurring in

all three sections is kept and saved as a point of the response surface associated with the particular member under consideration. The output obtained at the end of running the subroutine of program1.m dedicated to the calculation of the response surfaces of the individual members is again a 3D array, in which each face represents the response of the member within the member list with index coinciding with the third dimension of the array. The location and name of the array are input on page six.

6. MODELING OF THE WIND DIRECTIONALITY EFFECTS AND CALCULATION OF THE MRIS (PROGRAM2.MAT)

For design purposes it is necessary to ascertain the peak demand for the various global and local responses of interest corresponding to specified MRIs. This can be achieved by following the procedure described in [17, 18], which makes use of a directional extreme wind database such as the simulated hurricane wind speed database publicly available at www.nist.gov/wind. Work on a procedure for developing by simulation similar databases for non-hurricane winds is in progress, and is scheduled for publication and inclusion in the site www.nist.gov/wind in early 2009.

To explain the method consider having $p=1,2...,p_{max}$ storms, each of which contains the maximum hourly wind speed for a certain number of directions $\gamma=1,2,....,36$. From the peak response surfaces it is possible to calculate the peak response of each constraint for every direction and storm $[d_{jks}^{peak}]_{\gamma p}$. From a design point of view, however, it is the maximum value that occurs during the storm that is of interest. Therefore only the quantities $d_{jksp} = \max_\gamma \left\{[d_{jks}^{peak}]_{\gamma p}\right\}$ are considered. This means that for each storm there is a single peak response for every constraint. The set of quantities $\{d_{jksp}\}$ for $p=1,2...,p_{max}$ represent an univariate sample of size p_{max}. By rank-ordering the set $\{d_{jksp}\}$ it is possible to estimate the peak response for a given MRI. If the rate of arrival of storms for the site being considered is ν, then the m-th largest response of interest will have a MRI given by

$$MRI = \frac{p_{max}+1}{\nu m}$$ (23)

Therefore by recording the wind direction and mean wind speed at the top of the building for every data point in the set $\{d_{jksp}\}$ the direction and mean wind speed can be obtained for any given MRI and constraint.

The execution of program2.m will initiate a run through all the storm events contained in the database. The wind directions WD and wind speeds WS for which the structural response has been calculated during the execution of program1.m do not necessarily coincide with those contained in the hurricane database. To overcome this problem two interpolation schemes are available. Method A interpolates linearly between wind speeds and wind directions. Method B interpolates linearly between wind speeds, but considers only the larger of the two values corresponding to the two bounding wind directions. Method B is slightly more conservative.

The output files from program2.m depend on the response under consideration. For the structural members a matrix is output that contains along each row the demand corresponding to the the MRIs to be estimated. In the case of the inter-story drift a 3D array is used to store the information. Each face of this array contains the x and y demands corresponding to the MRI contained in the vector of MRIs input specifically for the inter-story drift and top floor acceleration. A separate file is generated for every column line input. The demands corresponding to the top floor accelerations are saved to a matrix that contains the x and y acceleration demands as columns, while each row corresponds to a particular MRI contained in the relevant vector. Again a separate file is saved for each point considered.

7. VERIFICATION OF THE HR_DAD_1.1 SOFTWARE

7.1. Introduction

In order to verify the proper functioning of the HR_DAD_1.1 software a simple example will be solved by HR_DAD_1.1 and compared to the closed form solution. The example represents a simple 2D 2-story building, assumed to be linearly elastic and have negligible P-Δ effects. The solution of the example can be found in [20, Chapter 12], which contains additional details.

7.2. 2D 2-story building

7.2.1. Closed form solution

The building is shown in Fig. 7.1. The beams are considered to be rigid. Various responses will be calculated considering the time dependent load $p \sin \omega t$ applied to the first floor and compared to those obtained from HR_DAD_1.1. The responses that will be compared are: the time histories of the floor displacements u_1 and u_2; the peak observed inter-story drift and second floor peak acceleration; the standard deviation of the peak bending moment occurring in the columns C1, C2, C3 and C4 (Fig. 7.1).

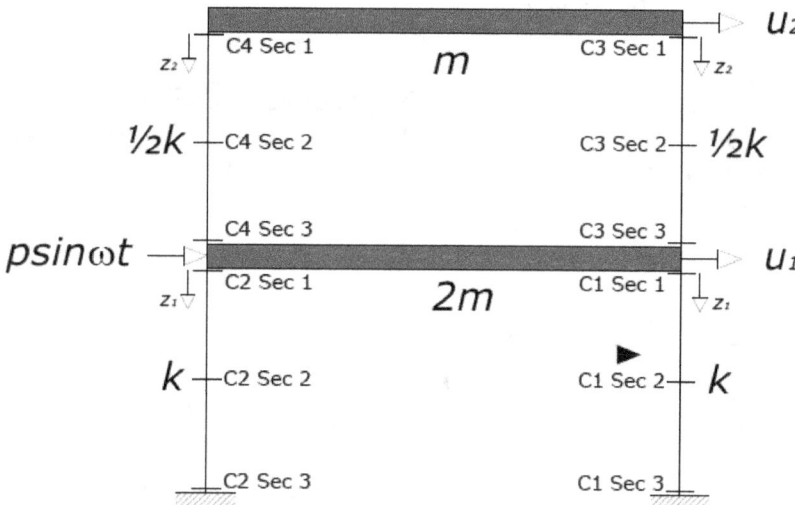

Figure 7.1. 2D 2-story example.

The closed form solution for the displacements can be found in [20]:

$$u_1(t) = \frac{p}{6k}\left[(2C_1 + C_2)\sin \omega t + (2D_1 + D_2)\cos \omega t\right] \tag{24a}$$

$$u_2(t) = \frac{p}{6k}\left[(4C_1 - C_2)\sin \omega t + (4D_1 - D_2)\cos \omega t\right] \tag{24b}$$

where:

$$C_n = \frac{1-(\omega/\omega_n)^2}{[1-(\omega/\omega_n)^2]^2 + (2\zeta_n\,\omega/\omega_n)^2} \quad \text{for } n = 1, 2 \tag{25a}$$

15

$$D_n = \frac{-2\zeta_n \omega/\omega_n}{[1-(\omega/\omega_n)^2]^2 + (2\zeta_n \omega/\omega_n)^2} \quad \text{for } n = 1,2 \qquad (25b)$$

where $\omega_1 = \sqrt{k/2m}$ while $\omega_2 = \sqrt{2k/m}$. While ζ_n is the damping ratio of the nth vibrational mode.

By definition the time histories of the bending moments accruing in columns C1, C2, C3 and C4 will be given respectively by:

$$M_{C1-C2}(z_1, t) = \left[(z_1 - 2)k \right] u_1(t) \qquad (26a)$$

$$M_{C3-C4}(z_2, t) = \left[\left(\frac{z_2 - 2}{2} \right) k \right] (u_2(t) - u_1(t)) \qquad (26b)$$

Due to the nature of the structure and loading, the maximum bending moment will occur in the initial or terminal sections of the columns. Therefore, by choosing the initial section the standard deviation will be given by:

$$\sigma_{M_{C1-C2}} = \sqrt{E\left[\left(-2ku_1(t) - E(-2ku_1(t)) \right)^2 \right]} \qquad (27a)$$

$$\sigma_{M_{C3-C4}} = \sqrt{E\left[\left(-k(u_2(t) - u_1(t)) - E(-k(u_2(t) - u_1(t))) \right)^2 \right]} \qquad (27b)$$

where E is the expectation operator.

The maximum observed inter-story drift will be given by [20]:

$$\frac{Max[u_1(t)]}{h} = \frac{p}{6hk} \sqrt{(2C_1 + C_2)^2 + (2D_1 + D_2)^2} \qquad (28a)$$

$$\frac{Max[u_2(t) - u_1(t)]}{h} = \frac{p}{6hk} \sqrt{(2C_1 - 2C_2)^2 + (2D_1 - 2D_2)^2} \qquad (28b)$$

where h is the story height.

The maximum observed top floor acceleration is given by:

$$Max\left[\frac{d^2 u_2(t)}{dt^2} \right] = \frac{\omega^2 p}{6k} \sqrt{(C_2 - 4C_1)^2 + (D_2 - 4D_1)^2} \qquad (29)$$

For the worked example the values for the various parameters are listed in Table 7.1.

16

Table 7.1. Summary of input for running the example (closed form solution)

Mechanical and geometrical parameters
Story height h = 4m
Modal damping ζ_n = 1.5% of critical for n = 1,2
Stiffness k = 400000 N/m
Mass m = 15000 kg
p = 10000 N
ω = 1 rad/s

7.2.2 HR_DAD_1.1 solution

HR_DAD_1.1 is developed for analyzing 3D structures. However it is easy to consider a 2D structure by simply entering as input the relative information for the plane in which it has been decided to analyze the response. In this example the xz plane is chosen. The forcing function $p \sin \omega t$ is described considering a sampling frequency of 20 Hz for an observation time of 1600 s and is loaded on page 2 of the HR_DAD_1.1 GUI as a mat file which also contains the time histories of the loads applied to the other degrees of freedom, which are equal to zero. The input necessary to run HR_DAD_1.1 is of two types. The first type is directly entered from the GUI while the second form of input consists of Matlab mat files, e.g., for the forcing function, also loaded through the GUI. Tables 7.2 and 7.3 summarize the input necessary for running the example. The variables are organized by their respective pages within the program. The definition of the variables can be found within the pages of the software and variables not applicable to this particular wind effect of interest are omitted. The eigenvectors input by the variable *evectors* (Table 7.3, page 1) come from [20]. Table 7.4 shows the influence coefficients arranged in their 3D array for the 4 columns of interest.

Table 7.2. Summary of type 1 HR_DAD input for running the example

Page 1	Page 2	Page 3	Page 4
Nfloors = 2	Vm = 1	WS = 1	DLf = 1
H_bldg = 8	freq = 20	WD = 0	SDLf = 1
F_dofs = 3	ms = 1		LLf = 1
Modes = 6	Npoints = 32001		WLf = 1
Modal periods = [1.7207 1 1 0.8604 1 1]	Nstart = 2000		g = 1
Modal damping = [1.5 1 1 1.5 1 1]			

Table 7.3. Summary of type 2 HR_DAD input for running the example

Page 1	Page 3	Page 4

$$evectors = \begin{bmatrix} 0.5 & 0 & 0 & -1 & 0 & 0 \\ 1 & 0 & 0 & 1 & 0 & 0 \\ 0 & 1 & 0 & 0 & -1 & 0 \\ 0 & 1 & 0 & 0 & 1 & 0 \\ 0 & 0 & 1 & 0 & 0 & -1 \\ 0 & 0 & 1 & 0 & 0 & 1 \end{bmatrix}$$

$$props = \begin{bmatrix} 1 & 1 & 1 & 1.111 & 1 \\ 2 & 1 & 1 & 1.111 & 1 \\ 3 & 1 & 1 & 1.111 & 1 \\ 4 & 1 & 1 & 1.111 & 1 \end{bmatrix}$$

$$mem\_list = \begin{bmatrix} 1 & 2 & 3 & 4 \\ 'C' & 'C' & 'C' & 'C' \end{bmatrix}$$

$$mass = \begin{bmatrix} 30000 \\ 1 \\ 1 \\ 15000 \\ 1 \\ 1 \end{bmatrix}$$

Page 4	Page 5

$$frames\_DL = \begin{bmatrix} 1 & 0 & 0 & 0 & 0 & 0 & 0 & 0 & 0 & 0 \\ 2 & 0 & 0 & 0 & 0 & 0 & 0 & 0 & 0 & 0 \\ 3 & 0 & 0 & 0 & 0 & 0 & 0 & 0 & 0 & 0 \\ 4 & 0 & 0 & 0 & 0 & 0 & 0 & 0 & 0 & 0 \end{bmatrix}$$

$$frames\_SDL = \begin{bmatrix} 1 & 0 & 0 & 0 & 0 & 0 & 0 & 0 & 0 & 0 \\ 2 & 0 & 0 & 0 & 0 & 0 & 0 & 0 & 0 & 0 \\ 3 & 0 & 0 & 0 & 0 & 0 & 0 & 0 & 0 & 0 \\ 4 & 0 & 0 & 0 & 0 & 0 & 0 & 0 & 0 & 0 \end{bmatrix}$$

$$frames\_LL = \begin{bmatrix} 1 & 0 & 0 & 0 & 0 & 0 & 0 & 0 & 0 & 0 \\ 2 & 0 & 0 & 0 & 0 & 0 & 0 & 0 & 0 & 0 \\ 3 & 0 & 0 & 0 & 0 & 0 & 0 & 0 & 0 & 0 \\ 4 & 0 & 0 & 0 & 0 & 0 & 0 & 0 & 0 & 0 \end{bmatrix}$$

$$interstory\_location = \begin{bmatrix} 0 & 0 & 4 \\ 0 & 0 & 4 \end{bmatrix}$$

$$acceleration\_location = \begin{bmatrix} 0 & 0 \end{bmatrix}$$

Page 7

$$displacement\_location = \begin{bmatrix} 0 & 0 \\ 0 & 0 \end{bmatrix}$$

Table 7.4. Influence coefficients for running the example

Page 3	
Elements 1 and 2	Elements 3 and 4

$$dif(:,:,1) = dif(:,:,2) = \begin{bmatrix} 0 & 0 & 0 & 0 & -1 & 0 \\ 0 & 0 & 0 & 0 & -1 & 0 \\ 0 & 0 & 0 & 0 & 0 & 0 \\ 0 & 0 & 0 & 0 & 0 & 0 \\ 0 & 0 & 0 & 0 & 0 & 0 \\ 0 & 0 & 0 & 0 & 0 & 0 \\ 0 & 0 & 0 & 0 & 0 & 0 \\ 0 & 0 & 0 & 0 & 0 & 0 \\ 0 & 0 & 0 & 0 & 0 & 0 \\ 0 & 0 & 0 & 0 & 0 & 0 \\ 0 & 0 & 0 & 0 & 0 & 0 \\ 0 & 0 & 0 & 0 & 0 & 0 \\ 0 & 0 & 0 & 0 & 1 & 0 \\ 0 & 0 & 0 & 0 & 1 & 0 \\ 0 & 0 & 0 & 0 & 0 & 0 \\ 0 & 0 & 0 & 0 & 0 & 0 \\ 0 & 0 & 0 & 0 & 0 & 0 \\ 0 & 0 & 0 & 0 & 0 & 0 \end{bmatrix} \qquad dif(:,:,3) = dif(:,:,4) = \begin{bmatrix} 0 & 0 & 0 & 0 & 0 & 0 \\ 0 & 0 & 0 & 0 & -1 & 0 \\ 0 & 0 & 0 & 0 & 0 & 0 \\ 0 & 0 & 0 & 0 & 0 & 0 \\ 0 & 0 & 0 & 0 & 0 & 0 \\ 0 & 0 & 0 & 0 & 0 & 0 \\ 0 & 0 & 0 & 0 & 0 & 0 \\ 0 & 0 & 0 & 0 & 0 & 0 \\ 0 & 0 & 0 & 0 & 0 & 0 \\ 0 & 0 & 0 & 0 & 0 & 0 \\ 0 & 0 & 0 & 0 & 0 & 0 \\ 0 & 0 & 0 & 0 & 0 & 0 \\ 0 & 0 & 0 & 0 & 0 & 0 \\ 0 & 0 & 0 & 0 & 1 & 0 \\ 0 & 0 & 0 & 0 & 0 & 0 \\ 0 & 0 & 0 & 0 & 0 & 0 \\ 0 & 0 & 0 & 0 & 0 & 0 \\ 0 & 0 & 0 & 0 & 0 & 0 \end{bmatrix}$$

HR_DAD_1.1 will calculate the element response in the initial, mid and terminal sections as shown in Fig. 7.1. The maximum response will then be saved to the output file. In the design of tall buildings it is not the individual internal forces that are of interest, but rather their combination defined by certain interaction formulas like Eqs. H1-1a and H1-1b contained in the American Institute of Steel Construction [21] manual for steel structures. Due to the Gaussian nature of the local member-based response of tall buildings, the peaks of the interaction formulas can be calculated using the Rice formula. This hypothesis is obviously not valid for this simple example. However the purpose of this example is not the accurate calculation of its response, but rather the validation of the calculations being performed by HR_DAD_1.1. Therefore by specifying an input value of 1.111 Nm for the nominal flexural strengths of the columns and by ignoring the axial forces in the various columns, the output of HR_DAD_1.1 should yield the following parameter:

$$b_i = \max_{Sec1,Sec2,Sec3} \left[\overline{M_i} + \sigma_{M_i} \right] \text{ for } i = 1,2,...,4 \qquad (30)$$

Therefore by considering the particular geometry and loading of the structure, $b_i - \overline{M}_i$(sec1 or sec3) for $i = 1, 2, ..., 4$, should coincide with the standard deviation of the bending moment in member i calculated in section 1 or 3 from Eqs. 27a-b.

Once the relative files and variables are defined and loaded, the solution maybe calculated by simply invoking program1.m by pressing RUN on page 6. The solution will be saved to specified file locations and names input on pages 6 and 7 of the GUI.

7.2.3. Comparison between closed form and numerical results

Figures 7.2 and 7.3 show the comparison between the time histories of the displacements u_1 and u_2 obtained in closed form from Eqs. 24a-b, with those estimated by running HR_DAD_1.1. As can be seen the numerical results are in good agreement with the exact solution.

Table 7.5 contains the comparison of various response parameters. Good agreement can be seen in all cases with any differences due to the numerical integration.

7.3. Summary

The purpose of this exercise was to verify the functioning of the HR_DAD software package by comparing the closed form output from a simple 2D two story shear frame with those obtainable from running HR_DAD. It was seen that HR_DAD does indeed accurately estimate the response components necessary for the design of framed structures.

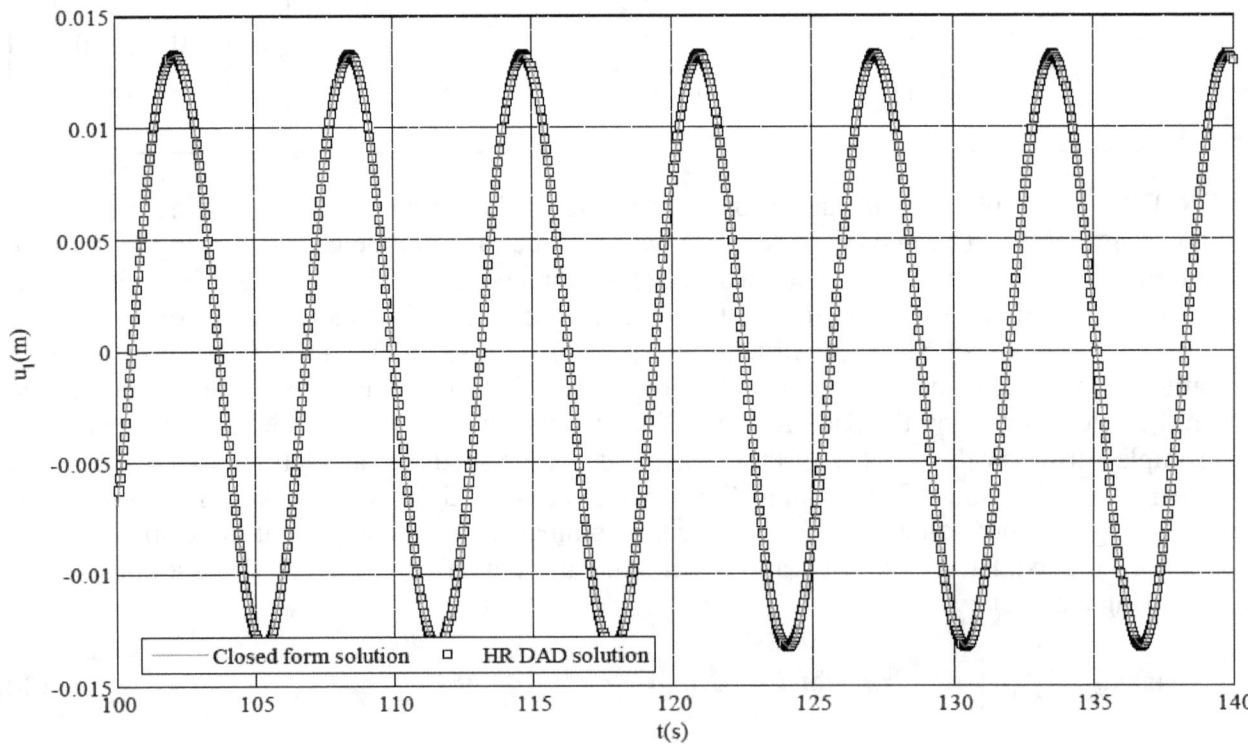

Figure 7.2. Comparison of closed form and HR_DAD time histories of u_1

20

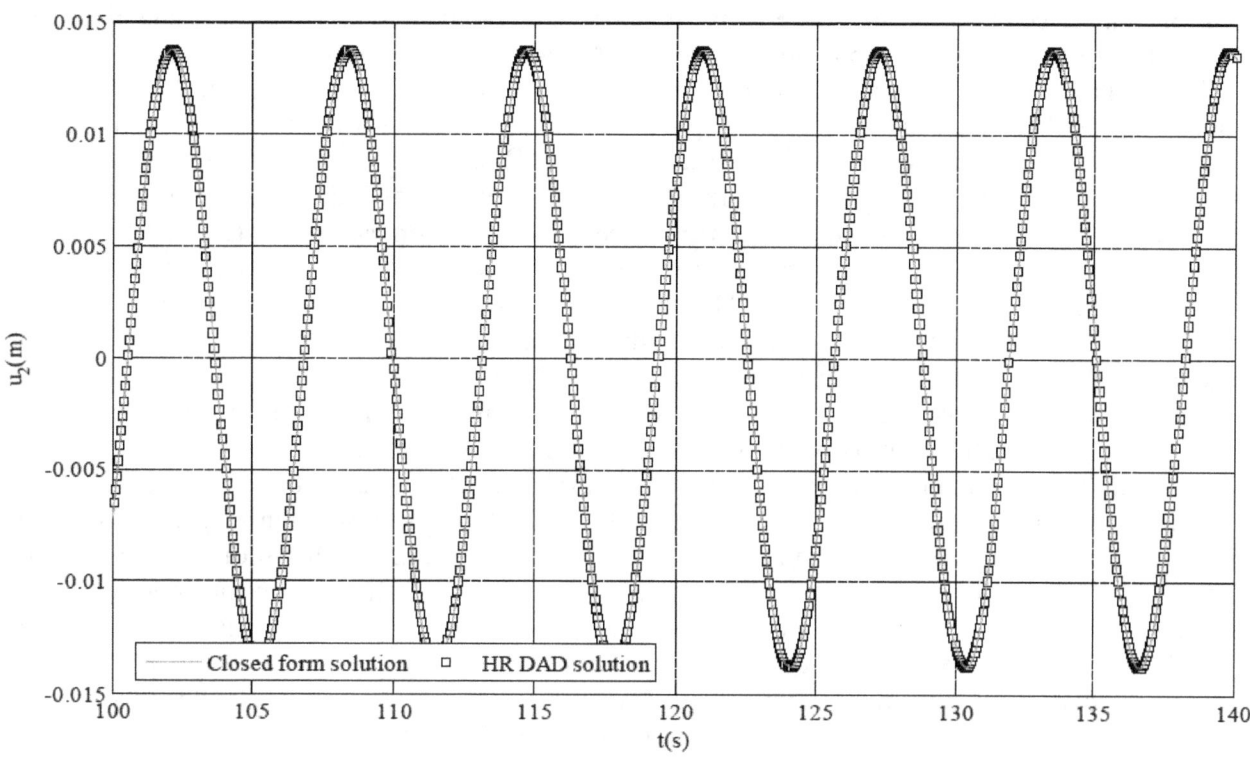

Figure 7.3. Comparison of closed form and HR_DAD time histories of u_2

Table 7.5. Comparison of various response components

Response parameter	Closed form solution	HR DAD
Max σ_M		
column 1	7495.8	7494.5
column 2	7495.8	7494.5
column 3	146.4060	145.9744
column 4	146.4060	145.9744
Max inter-story drift		
floor 1	0.0033	0.0033
floor 2	0.00013	0.00013
Max second floor acceleration	0.0138	0.0139

8. CONCLUDING REMARKS

This report describes the enhancements that have been implemented in HR_DAD in order to make this a realistic and efficient package for the dynamic time domain structural analysis of tall buildings subjected to windstorm events. The inherent advantages of working in the time domain include (1) capability of directly accounting for nonlinear and/or 3D coupled fundamental modal shapes; (2) the possibility of considering any number of modes in the response; (3) the capability of directly estimating the correlation that exits between the modal responses; (4) the possibility to include non-

Gaussian effects; (5) the preservation of all phase relations, which avoids the need to consider more than one single combination of load effects, as opposed to the numerous load effect combinations required if the frequency domain approach is used. A possible drawback of a time domain dynamic analysis approach is the long computational time necessary compared to more traditional, frequency domain approaches. Due to the extremely large number of members the processes of calculating the peak demand for each cross section for a specified MRI could become prohibitive. However, this is avoided by considering the local response – the response of individual members - to be Gaussian. The peak response for each wind speed and direction is then calculated from the covariance matrix of the time-varying global floor loads, thus avoiding the evaluation of time series of individual internal forces. The global demands, inter-story drift and top floor acceleration are only on the order of hundreds, therefore allowing the peak response to be calculated directly from the time series and, therefore, the inclusion of non-Gaussian effects. This approach to the evaluation of the peak response has allowed HR_DAD_1.1 to become an efficient and competitive alternative to traditional methods of analysis.

This report contains the results from the completion of the first objective of a far more ambitious plan. Indeed, it is the first step toward the definition and resolution of a member size optimization problem formulated in the time domain for the design of tall buildings subjected to both local strength requirements and global inter-story drift constraints. Due to the enormous size of this optimization problem, which entails tens of thousands of non-linear constraints, the use of generic optimizers is out of the question. The only way to efficiently solve the problem is through specifically designed algorithms. The most appropriate problem definition and resolution algorithms have been the subject of research carried out in parallel with the continued development of HR_DAD_1.1. Details can be found in [19]. At present these algorithms are in the process of being implemented in the HR_DAD_1.1 environment and should be available in the next version of the software.

REFERENCES

[1] Kareem A., Dynamic Response of High-Rise Buildings to Stochastic Wind, *Journal of Wind Engineering and Industrial Aerodynamics*, 41-44, pp. 1101-1112, 1992.

[2] Chen X., Kareem A., Dynamic Wind Effects on Buildings with 3D Coupled Modes: Application of High Frequency Force Balance Measurements, *ASCE Journal of Engineering Mechanics*. Vol. 131, No. 11, pp.1115–1125, 2005.

[3] Spence S.M.J., Gioffrè M., Gusella V., Higher mode contributions to the response of tall buildings with regular and irregular profiles, *10° Convegno Nazionale di Ingegneria del Vento*, IN-VENTO-2008, Cefalù (PA), 8-11 Giugno 2008.

[4] Spence S.M.J., Gioffrè M., Gusella V., Influence of higher modes on the dynamic response of irregular and regular tall buildings, *6th International Colloquium on Bluff Bodies Aerodynamics and Applications* (BBAA VI), Milan, Italy, July 20-24, 2008.

[5] P.J. Vickery, A.C. Steckley, N. Isyumov, B.J Vickery. The effect of mode shape on the wind induced response of tall buildings, *Proceedings of the 5th United States National Conference on Wind Engineering*, Lubbock, Texas, 1B-41-1B-48, 1985.

[6] Boggs, D. W., and Peterka, J. A. Aerodynamic model tests of tall buildings. *ASCE Journal of Engineering Mechanics,* Vol.115 N. 3, pp. 618-635, 1989.

[7] Xu, Y. L., and Kwok, K. C. S. Mode shape corrections for wind tunnel tests of tall buildings. *Engineering Structures*, Vol. 15, pp. 618-635, 1993.

[8] Zhou, Y., Kareem, A., and Gu, M. Mode shape corrections for wind load effects. *ASCE Journal of Engineering Mechanics,* Vol. 128 N. 1, pp. 15-23, 2002.

[9] Holmes, J. D., Rofail, A., and Aurelius, L. High frequency base balance methodologies for tall buildings with torsional and coupled resonant modes. *Proc., 11th International Conf. on Wind Engineering*, Texas Tech Univ., Lubbock, Tex., pp. 2381-2388, 2003.

[10] Chen, X., and Kareem, A. Coupled building response analysis using HFFB: Some new insights. *Proc., 5th Bluff Body Aerodynamics and Applications (BBAAV)*, Ottawa, 2004.

[11] Ohkuma, T., Marukawa, H., Yoshie, K., Niwa, H., Teramoto, T., and Kitamura, H. Simulation method of simultaneous time-series of multi-local wind forces on tall buildings by using dynamic balance data. *Journal of Wind Engineering and Industrial Aerodynamics*, 54-55, pp. 115-123, 1995.

[12] Yip, D. Y. N., and Flay, R. G. J. A new force balance data analysis method for wind response predictions of tall buildings. *Journal of Wind Engineering and Industrial Aerodynamics*, 54-55, 457–471, 1995.

[13] Solari, G., Reinhold, T. A., and Livesey, F. "Investigation of wind actions and effects on the leaning Tower of Pisa." *Wind & Structures*, 1-1, pp.1-23, 1998.

[14] Chen, X., and Kareem, A. Validity wind load distribution based on high frequency force balance measurements. *ASCE Journal of Structural Engineering*, 131 no. 6, 984-987, 2005.

[15] Xie J., Progress of Wind Tunnel Techniques in Practical Applications, *The 4th International Conference on Advances in Wind and Structures* AWAS'08, Jeju, Korea, 29-31, May, 2008.

[16] Main,J. A. and Fritz, W. A., Database-Assisted Design for Wind: Concepts, Software, and Examples for Rigid and Flexible Buildings, NIST Building Science Series 180, 2006.

[17] Simiu, E. and Miyata, T., *Design of Buildings and Bridges for Wind.* Wiley, Hoboken, New Jersey, 2006.

[18] Simiu, E., Gabbai, R.D. and Fritz, W.P.,Wind-induced tall building response: a time-domain approach. *Wind and Structures,* Vol. 11, No. 6, 427-440, 2008.

[19] Spence S.M.J., Gioffrè M., A Database-Assisted Design approach for lateral drift optimization of tall buildings, *14th IFIP WG 7.5 Working Conference on Reliability and Optimization of Structural Systems (IFIP 08-WG 7.5)*, Mexico City, Mexico, August 6-9, 2008.

[20] Chopra A. K. Dynamics of structures: Theory and applications to earthquake engineering. *2nd ed. Prentice-Hall*; 2000.

[21] American National Standard ANSI/AISC 360-05, Specification for Structural Steel Buildings. *American Institute of Steel Construction, Inc.* Chicago, Illinois; 2005.

Appendix

User's Manual

High-Rise Database-Assisted Design 1.1 (HR_DAD_1.1)
Software

Ssoftware available at www.nist.gov/wind

Disclaimer: Certain trade names or company products are specified in this document to specify adequately the procedure used. Such identification does not imply recommendation or endorsement by NIST, nor does it imply that the product is the best available for the purpose. The "stand-alone" version of this software requires installation of the MATLAB[1] Component Runtime (MCR) Libraries provided by The MathWorks, Inc. The author's limited rights to the deployment of this program are limited by a license agreement between NIST and The MathWorks. The license agreement can be found at www.mathworks.com/license/. The author, NIST, and The MathWorks and its licensors are excluded from all liability for damages or any obligation to provide remedial actions.

---

[1]MATLAB®. © 1984 – 2005 The Mathworks, Inc.

A1. Introduction

This User's Manual is designed to assist the user of the software HR_DAD_1.1. The software calculates the response of tall buildings subjected to wind loads, including internal forces in members, member interaction formulas based on demand-to-capacity ratios, inter-story drift, and accelerations, for any specified mean recurrence interval of the wind effect being considered.

A2. How to download and install the software

The **HR_DAD_1.1** software has been developed using the MATLAB® language and can be accessed via the internet site http://www.nist.gov/wind. Within the site, click the link "Wind Effects on Flexible Buildings". This opens the main page "HR_DAD_1.1 - DAD Software for High-Rise   Buildings"   (direct   access: http://www.itl.nist.gov/div898/winds/hr_dad_1.1/hr_dad_1.1.htm). The files available for download are all in the bulleted list under the heading *"Files Available for Download"*. In the following, reference is made to each set of files to be downloaded by the name of the associated bullet. From the list, first consider the "Files for **HR_DAD_1.1** software" bullet. Next to the title of the bullet item, there is a link to the self-extracting file zip file "HR_DAD_1.1.exe", which contains the stand-alone executable. To run HR_DAD_1.1 the MCRInstaller must first be installed by executing the application "MRCInstaller.exe". The HR_DAD_1.1 software can then be launched by double-clicking the file "HR_DAD.exe" within the "stand-alone" folder. This action opens the ten figure windows ('Page_Open', 'Page_Main', 'Page_One', 'Page_Two' … 'Page_Eight') that form the graphical user interface (GUI).

A3. Basics of using the **HR_DAD_1.1** software

The ten figure windows (i.e., the "pages") opened above are used primarily to (1) assign values to the variables used by the **HR_DAD_1.1** software ('Page_One' through 'Page_Five'), (2) to perform the calculations ('Page_Six'). Variable values can be assigned in any order in 'Page_One' through 'Page_Five'. The variable names within **HR_DAD_1.1** are typically shown in parentheses before the input box on each page. In several instances, a saved MATLAB mat file is opened within a page to load variables that contain vectors, matrices, or 3D arrays. Help

icons, ⑦, are located next to the input boxes for key variables. For a given variable, clicking on the associated help icon will open a separate window that provides information such as the variable name, the required variable size, a description of the variable and the specific organization of the variable's contents (for vector, matrix or 3D array variables).

A4. How to use the Manual

This manual is organized with the aim of guiding the user through the necessary steps in order to run **HR_DAD_1.1**. This is done by first defining a simple 3D 2-story building. The input necessary for running this example in **HR_DAD_1.1** is then schematically outlined, allowing the user to get an idea of the software's operation. Each page of the graphical user interface (GUI) is then described in detail, together with what the input would be in the case of the simple 3D 2-story building.

A5. Definition of the 3D 2-story building

The structure is depicted in Fig. A1. The floors are considered rigid and no rotation is allowed between the floor slabs and the 8 columns (members 1 to 8).

A6. Input for running HR_DAD_1.1

To run **HR_DAD_1.1** information is necessary concerning the structure, the wind climate at an open site at a reasonable distance from the building, and the relation between the directional speeds at 10 m elevation at that site and the corresponding directional wind speeds (including veering effects, see [17, p. 15], as follows:

1. Information related to the dynamic properties of the structure:

    - **Mass** and **mass moment of inertia** with respect to the center of mass, for each floor..
    - **Mode shapes, frequencies, and damping ratios**. The mode shapes must be referenced to the centers of mass of each floor.

2. Information for member demand-to-capacity ratio calculations:

    - **Influence coefficients,** defined as *the internal force that occurs in a given section of a given member due to a unit force applied to the center of mass of a given floor*.

3. Information relating to the structural loading. In particular **HR_DAD_1.1** is capable of considering:

- **Time varying loads** applied at the center of mass of each floor, obtained from wind tunnel tests carried out on rigid models
- **Static loads** applied to each member.

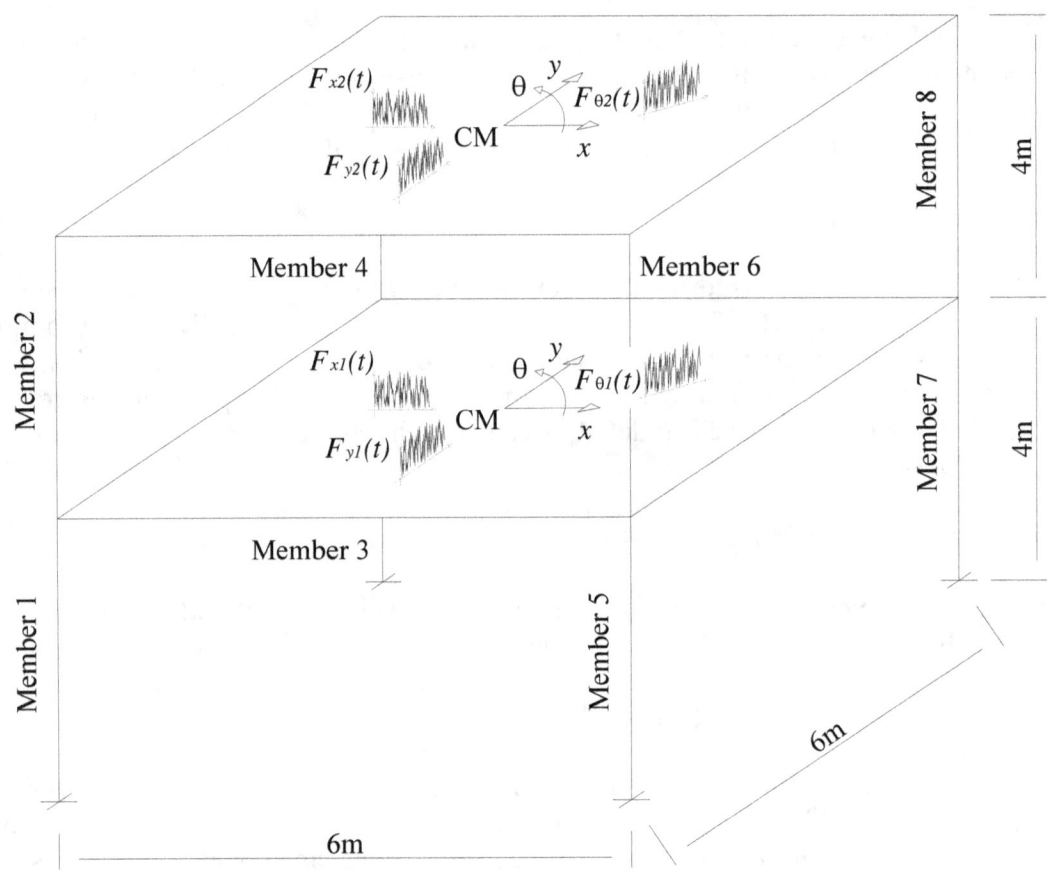

Figure A1.Simple 2- story 3D building.

4. Information relating to the wind climate at or near the building site

- In general, large separate **databases of directional wind speeds** are needed for each type of wind occurring at the site (e.g., **hurricanes, thunderstorms, synoptic winds**). The calculation of mean recurrence intervals of wind effects is analogous to the simple calculations discussed for wind speeds in [17, p. 28]. An algorithm for incorporating this calculation is currently being developed.

- Information regarding the ratio between wind speeds at 10 m above ground in open terrain and their mean hourly wind speed counterparts (as affected by veering, see

[17, p. 15] at the top of the building. An algorithm for incorporating the effect of veering is currently being developed.

A7. Output from Running HR_DAD_1.1

- **Peak demand/capacity ratios** for all members calculated for any number of Mean Recurrence Intervals (**MRIs**).
- **Peak inter-story drift** calculated for any number of column lines and **MRIs.**
- **Peak top floor acceleration** calculated for any number of points and any number of **MRIs.**

A8. Details on Use of the Software

The following pages will illustrate in detail how to use the eight pages that make up the GUI of **HR_DAD_1.1**. During the general description of the necessary input, and the formatting of such input, the variables necessary for running the example building shown in Fig. A1 will be input under the heading 3D 2-story example.

INPUT:

H_bldg = Height of the building in meters

      3D 2-story example: total building height is 8m.

Nfloors = Number of floors composing the building

      3D 2-story example: 2 floors.

Fdofs = Degrees of freedom per floor

      3D 2-story example: 3.

No. of modes = Choose the number of vibrational modes to be considered in the analysis

      3D 2-story example: 3 modes are chosen.

28

Modal Periods = assign the modal periods in seconds to each vibrational mode

3D 2-story example: if 3 modes are to be considered in the analysis and the periods are, for example, 1.7 s, 1.5 s and 0.8 s, then the following vector would be entered:

[0.6122 0.6102 0.4565]

Note: The variable must be input using square brackets as shown.

Modal damping, % = assign the modal damping as a percentage of critical.

3D 2-story example: if 3 modes are to be considered in the analysis and the modal damping is to 1.5% of critical for each mode, then the following vector should be entered;

[1.5 1.5 1.5]

Note: The variable must be input using square brackets as shown.

Mode shapes = Input the Matlab mat file containing the mode shapes.

File structure: The file can be constructed in MATLAB. The variable in the mat file containing the mode shapes must be named *evectors*. The file containing the variable maybe named anyway desired. The variable *evectors* is a matrix whose columns contain the mode shapes referred to the center of mass of each floor. Each column has the x-coordinates first then the y-coordinates and finally the θ-coordinates of the mode shape.

3D 2-story example: consider the following mode shapes:

$$\text{mode shape } 1 = \begin{bmatrix} 0.62 \\ 1 \\ 0 \\ 0 \\ 0 \\ 0 \end{bmatrix}, \text{ mode shape } 2 = \begin{bmatrix} 0 \\ 0 \\ 0.62 \\ 1 \\ 0 \\ 0 \end{bmatrix} \text{ and mode shape } 3 = \begin{bmatrix} 0 \\ 0 \\ 0 \\ 0 \\ 0.62 \\ 1 \end{bmatrix}.$$

The variable *evectors* would therefore take on the following form:

29

$$evectors = \begin{bmatrix} 0.62 & 0 & 0 \\ 1 & 0 & 0 \\ 0 & 0.62 & 0 \\ 0 & 1 & 0 \\ 0 & 0 & 0.62 \\ 0 & 0 & 1 \end{bmatrix}$$

and could be saved for example as ModeShapes.mat

END PAGE_ONE INPUT

30

INPUT:

flnFl = load the MATLAB mat file containing the time histories of the floor loads.

> *File structure*: The file can be constructed in MATLAB. The variable in the mat file containing the time histories must be named **F**. The mat file containing the variable **F** can then be saved under any name but must end with "**_XXX**" where **XXX** gives the wind direction, in degrees, from which the wind was blowing in the wind tunnel when the loads where ascertained. An example is shown in Fig. 2 for directions **XXX** = 000 and **XXX** = 035.
>
> For each wind direction a separate mat file is needed containing the relative variable **F**. The variable **F** is a matrix, each line of which contains the time history of a floor load acting in one of the directions x, y or θ. The first N (N is the number of floors) rows correspond to the floor loads acting in the x-direction starting from the first floor. The Next N rows correspond to the floor loads acting in the y-direction while the last N rows correspond to the floor loads acting in the θ-direction. **F** will have a total of $3N$ rows.

31

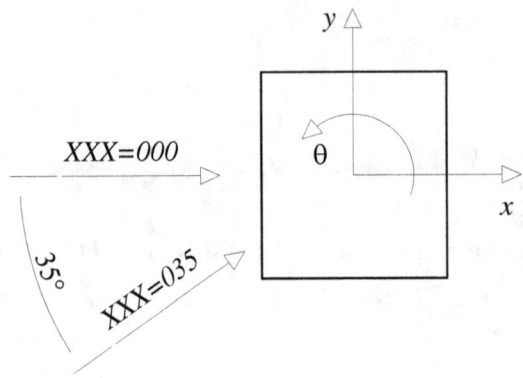

Figure A2.File naming convention used for the floor loads.

3D 2-story example: floor loads for running the example can be downloaded from the site www.nist.gov/wind.

Vm = mean wind speed in m/s at roof of the model during the wind tunnel tests.

3D 2-story example: 20m/s

freq = sampling frequency used in the wind tunnel.

3D 2-story example: 250Hz

ms = scale of the model used in the wind tunnel tests.

3D 2-story example: if the scale was 1/500 then the correct number to input is 500.

Npoints = total number of points that make up the time histories of the floor loads.

3D 2-story example: if 30s of data is recoded during the wind tunnel tests with a sampling frequency (**freq**) of 250Hz, then the time histories of the floor loads will have a total of 7504 points.

Nstart = number of points to be cut from beginning of the time histories during the analysis.

3D 2-story example: 200 points is reasonable.

Note: Numerical integration needs a certain number of points before it stabilizes. Therefore a certain number of initial points should be cut form the solution of the equations of motion before estimating the response parameters.

END PAGE_TWO INPUT

INPUT:

flnMem = load the list of members composing the structure.

> *File structure*: The file can be constructed in MATLAB. The variable in the mat file containing the list of members must be named **mem_list**. The mat file containing the variable **mem_list** can then be saved under any name. The variable **mem_list** is a matrix with two rows and number of columns equal to the number of members that make up the structure. The first row contains the member numbers while the second row contains a label identifying whether the member is a beam, column or diagonal. The labels must be a string of characters that begin with the letters "B" if the element is a Beam, "C" if the element is a column or "D" if the element is a diagonal.

> *3D 2-story example*: the structure has a total of 8 members. The variable **mem_list** would have the following appearance:

$$\textbf{mem\_list} = \begin{bmatrix} 1 & 2 & 3 & 4 & 5 & 6 & 7 & 8 \\ "C" & "C" & "C" & "C" & "C" & "C" & "C" & "C" \end{bmatrix}$$

and could be saved for example as members_list.mat

flnDif = load the mat file containing the influence coefficients.

File structure: The file can be constructed in MATLAB. The variable in the mat file containing the influence coefficients must be named **dif**. The mat file containing the variable **dif** can then be saved under any name. The variable **dif** is a 3D array. The following description of **dif** will be in reference to the member shown in Fig. 3.

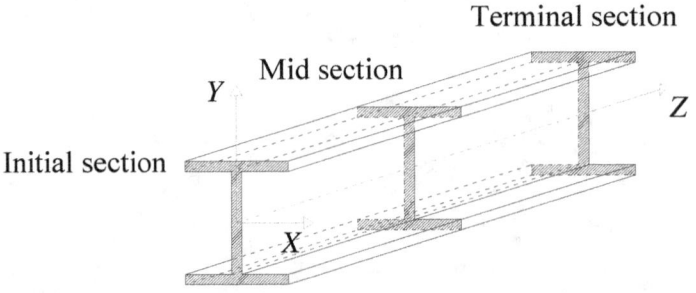

Figure A3. ith member of the structure.

Each face of **dif** contains the influence coefficients associated with the six internal forces (axial force, shear forces and bending moments in the local X and Y directions and torsion) that can arise in the initial, mid or terminal sections of a given member due to a unit force applied to the center of mass of a given floor in one of the directions x, y or θ. In particular, the first column contains the axial forces, second and third columns the X and Y shears, fourth column the torsion and the fifth and sixth columns the X and Y bending moments. The first N rows (N is the number of floors of the building) are the internal forces in the initial section due to unit forces applied to the centers of mass of each floor in the x-directions starting from the 1st floor. The next N rows are due to unit forces in the y-direction while successive N rows are due to unit torques applied to the centers of mass. This makes up a total of $3N$ rows for the influence coefficients of the initial section. The following $6N$ rows are the influence coefficients for the mid and terminal sections making a total of 9N rows. The index of each face identifies the member. Indeed the *i*th face *must* correspond to the member described in the *i*th column of the variable **mem_list**.

3D 2-story example: the structure in Fig. A1 has 8 members all of which are columns. Assume them to be steel tubular members with an outside dimension of 25cm and flange thickness of 6mm. Also assume the steel to have an elastic

34

modulus of 200,000 N/mm^2. The next paragraph will explain how to calculate the first face of the array **dif** for this example building.

The first face will coincide with the element in the first column of the variable **mem_list**. It will therefore coincide with member 1. Because the structure has 2 floors, the first face will have 18 rows. The easiest way to calculate the influence coefficients is by constructing a model of the building in a commercial software analysis program, e.g., SAP2000. Within this environment it is then possible to define $3N = 6$ (N is the number of floors which in this case is 2) load cases, one for each unit force applied to the centers of mass of each floor. By running the analysis for the 6 load cases and exporting the results in Microsoft Excel format, tables similar to the one shown in Table A1 are obtained, where the load cases have been named influx1, influx2,...,influz2. The influence coefficients are simply the internal forces highlighted in red. To create the MATLAB mat file simply read the relative rows and columns into MATLAB and construct the array.

Table A1. SAP2000 Excel spreadsheet output for member 1.

TABLE: Element Forces - Frames									
Frame	Station	Output Case	Case Type	P	V2	V3	T	M2	M3
Text	m	Text	Text	N	N	N	N-m	N-m	N-m
1	0	influx1	LinStatic	0	0.25	0	0	0	0.5
1	0	influx2	LinStatic	0	0.25	0	0	0	0.5
1	0	influy1	LinStatic	0	0	0.25	0	0.5	0
1	0	influy2	LinStatic	0	0	0.25	0	0.5	0
1	0	influz1	LinStatic	0	0.039	-0.039	0.011	-0.079	0.079
1	0	influz2	LinStatic	0	0.039	-0.039	0.011	-0.079	0.079
1	2	influx1	LinStatic	0	0.25	0	0	0	0
1	2	influx2	LinStatic	0	0.25	0	0	0	0
1	2	influy1	LinStatic	0	0	0.25	0	0	0
1	2	influy2	LinStatic	0	0	0.25	0	0	0
1	2	influz1	LinStatic	0	0.039	-0.039	0.011	0	0
1	2	influz2	LinStatic	0	0.039	-0.039	0.011	0	0
1	4	influx1	LinStatic	0	0.25	0	0	0	-0.5
1	4	influx2	LinStatic	0	0.25	0	0	0	-0.5
1	4	influy1	LinStatic	0	0	0.25	0	-0.5	0
1	4	influy2	LinStatic	0	0	0.25	0	-0.5	0
1	4	influz1	LinStatic	0	0.039	-0.039	0.011	0.079	-0.079
1	4	influz2	LinStatic	0	0.039	-0.039	0.011	0.079	-0.079

The mat file containing the variable **dif** can be saved, for example, as dif_all.mat

flnProps = load the mat file containing the section properties.

File structure: The file can be constructed in MATLAB. The variable in the mat file containing the list of members must be named **props**. The mat file containing the variable **props** can then be saved under any name. The variable **props** is a matrix with first column coinciding with the first row of **mem_list** and therefore contains the member numbering scheme. Each row contains the strength capacities of the member identified in the first column of the row. In particular the second column of each row contains the maximum tensile strength, the third column the maximum compressive strength while the forth and fifth columns the bending strengths in the two local X and Y directions.

3D 2-story example: The variable **props** has the following appearance:

$$\textbf{props} = \begin{bmatrix} 1 & 1453525 & 1308172 & 133024 & 133024 \\ 2 & 1453525 & 1308172 & 133024 & 133024 \\ 3 & 1453525 & 1308172 & 133024 & 133024 \\ 4 & 1453525 & 1308172 & 133024 & 133024 \\ 5 & 1453525 & 1308172 & 133024 & 133024 \\ 6 & 1453525 & 1308172 & 133024 & 133024 \\ 7 & 1453525 & 1308172 & 133024 & 133024 \\ 8 & 1453525 & 1308172 & 133024 & 133024 \end{bmatrix}$$

The mat file containing the variable **props** could be saved as frame_properties.mat

WS = wind speeds at roof height of the full scale building for which the response is desired.

3D 2-story example: for example, if 5m/s to 50m/s with 5m/s increments are desired then the following vector would be entered:

[5 10 15… 45 50]

Note: The variable must be input using square brackets as shown. Please note that the following notation maybe used [5:5:50].

WD = wind directions for which the response is desired.

3D 2-story example: for example, if 0°, 10°, 20°… 350° and 360° are desired, then the following vector would be entered:

[0 10 20…350 360]

Note 1: The angles input in WD must be the same as or a subset of the angles for which the wind loading is known.

Note 2: The variable must be input using square brackets as shown. Please note that the following notation may be used [0:10:360].

MRIBij = The Mean Recurrence Intervals for which the peak demand/capacity ratios are wanted.

> *3D 2-story example*: for example, if the peak demand/capacity ratios for Mean Recurrence Intervals of 20 and 50 years are desired, then the following vector would be entered:
>
> [20 50]
>
> *Note:* The variable must be input using square brackets as shown.

MRIDrAc = The Mean Recurrence Intervals for which the peak inter-story drift and top floor acceleration is desired.

> *3D 2-story example*: for example, if the peak inter-story drift and top floor acceleration for Mean Recurrence Intervals of 10 and 20 years are desired, then the following vector would be entered:
>
> [10 20]
>
> *Note:* The variable must be input using square brackets as shown.

intmeth = choose the interpolation method that will be implemented during the calculation of the responses with given Mean Recurrence Intervals.

> *Note:* The methods will give similar results. However method A will give more accurate results but take longer than method B, which tends to be more conservative.

END PAGE_THREE INPUT

INPUT:

flnMass = load the mass and moments of inertia associated with each floor.

> *File structure*: The file can be constructed in MATLAB. The variable in the mat file containing the mass and moments of inertia must be named **mass**. The mat file containing the variable **mass** can then be saved under any name. The variable **mass** is a column vector. The first three rows are associated with the first floor. In particular the first entry is the mass in the *x*-direction, the second entry is the mass in the *y*-direction, while the third entry is the moment of inertia with respect to the center of mass. The next 3 rows are associated with the second floor and so forth.

> *3D 2-story example*: considering for both floors a mass in the *x* and *y* directions of 30000kg and moment of inertia of 310000kg m^2, the variable **mass** is:

$$\textbf{mass} = \begin{bmatrix} 30000 \\ 30000 \\ 310000 \\ 30000 \\ 30000 \\ 310000 \end{bmatrix}$$

The mat file containing the variable **mass** could be saved as mass_asc.mat

flnDLr = load file containing the contribution of the dead weight.

File structure: The file can be constructed in MATLAB. The variable in the mat file containing the contribution of the dead weight must be named **frames_DL**. The mat file containing the variable **frames_DL** can then be saved under any name. The variable **frames_DL** is a matrix with first column coinciding with the first row of **mem_list** and therefore contains the member numbering scheme. Each row contains the internal forces due to dead weight occurring in the initial, mid and terminal sections of the member identified in the first column of the row. In particular columns 2 to 4 contain the axial force and bending moments in the two local *X* and *Y* directions (Fig. A3) for the initial section. Columns 5 to 7 contain the same information for the mid section, while columns 8 to 10 are dedicated to axial and bending forces occurring in the terminal section.

3D 2-story example: from a SAP2000 or equivalent model of the building it is possible to access the internal forces that occur in the sections of interest of the 8 members composing the structure. This will give the following **frames_DL**:

$$\textbf{frames\_DL} = \begin{bmatrix} 1 & -153606 & 0 & 0 & -152704 & 0 & 0 & -151803 & 0 & 0 \\ 2 & -76803 & 0 & 0 & -75901 & 0 & 0 & -75000 & 0 & 0 \\ 3 & -153606 & 0 & 0 & -152704 & 0 & 0 & -151803 & 0 & 0 \\ 4 & -76803 & 0 & 0 & -75901 & 0 & 0 & -75000 & 0 & 0 \\ 5 & -153606 & 0 & 0 & -152704 & 0 & 0 & -151803 & 0 & 0 \\ 6 & -76803 & 0 & 0 & -75901 & 0 & 0 & -75000 & 0 & 0 \\ 7 & -153606 & 0 & 0 & -152704 & 0 & 0 & -151803 & 0 & 0 \\ 8 & -76803 & 0 & 0 & -75901 & 0 & 0 & -75000 & 0 & 0 \end{bmatrix}$$

The mat file containing the variable **frames_DL** could be saved as frames_DeadLoad.mat

flnLLr = load file containing the contribution of the live loads.

File structure: The file can be constructed in MATLAB. The variable in the mat file containing the contribution of the live loads must be named **frames_LL**. The mat file containing the variable **frames_SDL** can then be saved under any name. The variable **frames_SDL** is a matrix with first column coinciding with the first row of **mem_list** and therefore contains the member numbering scheme. Each row contains the internal forces due to superimposed dead load occurring in the initial, mid and terminal sections of the member identified in the first column of the row. In particular, columns 2 to 4 contain the axial force and bending moments in the two local X and Y directions (Fig. A3) for the initial section. Columns 5 to 7 contain the same information for the mid section, while columns 8 to 10 are dedicated to axial and bending forces occurring in the terminal section.

3D 2-story example: considering as superimposed dead load a distributed load of $0.5KN/m^2$ from the SAP2000 or equivalent model of the building it is possible to access the internal forces that occur in the sections of interest of the 8 members composing the structure. This will give the following **frames_SDL**:

$$
\textbf{frames\_SDL} =
\begin{bmatrix}
1 & -9000 & 0 & 0 & -9000 & 0 & 0 & -9000 & 0 & 0 \\
2 & -4500 & 0 & 0 & -4500 & 0 & 0 & -4500 & 0 & 0 \\
3 & -9000 & 0 & 0 & -9000 & 0 & 0 & -9000 & 0 & 0 \\
4 & -4500 & 0 & 0 & -4500 & 0 & 0 & -4500 & 0 & 0 \\
5 & -9000 & 0 & 0 & -9000 & 0 & 0 & -9000 & 0 & 0 \\
6 & -4500 & 0 & 0 & -4500 & 0 & 0 & -4500 & 0 & 0 \\
7 & -9000 & 0 & 0 & -9000 & 0 & 0 & -9000 & 0 & 0 \\
8 & -4500 & 0 & 0 & -4500 & 0 & 0 & -4500 & 0 & 0
\end{bmatrix}
$$

The mat file containing the variable **frames_SDL** could be saved as frames_SDeadLoad.mat

flnSDLr = load file containing the contribution of the superimposed dead load.

File structure: The file can be constructed in MATLAB. The variable in the mat file containing the contribution of the superimposed dead load must be named **frames_LL**. The mat file containing the variable **frames_LL** can then be saved under any name. The variable **frames_LL** is a matrix with first column coinciding with the first row of **mem_list** and therefore contains the member numbering scheme. Each row contains the internal forces due to live loads occurring in the initial, mid and terminal sections of the member identified in the first column of the row. In particular, columns 2 to 4 contain the axial force and bending moments in the two local X and Y directions (Fig. A3) for the initial section. Columns 5 to 7 contain the same information for the mid section while columns 8 to 10 are dedicated to axial and bending forces occurring in the terminal section.

3D 2-story example: considering as live load a distributed load of $2KN/m^2$ from the SAP2000 or equivalent model of the building it is possible to access the internal forces that occur in the sections of interest of the 8 members composing the structure. This will give the following **frames_LL**:

$$
\mathbf{frames\_LL} =
\begin{bmatrix}
1 & -36000 & 0 & 0 & -36000 & 0 & 0 & -36000 & 0 & 0 \\
2 & -18000 & 0 & 0 & -18000 & 0 & 0 & -18000 & 0 & 0 \\
3 & -36000 & 0 & 0 & -36000 & 0 & 0 & -36000 & 0 & 0 \\
4 & -18000 & 0 & 0 & -18000 & 0 & 0 & -18000 & 0 & 0 \\
5 & -36000 & 0 & 0 & -36000 & 0 & 0 & -36000 & 0 & 0 \\
6 & -18000 & 0 & 0 & -18000 & 0 & 0 & -18000 & 0 & 0 \\
7 & -36000 & 0 & 0 & -36000 & 0 & 0 & -36000 & 0 & 0 \\
8 & -18000 & 0 & 0 & -18000 & 0 & 0 & -18000 & 0 & 0
\end{bmatrix}
$$

The mat file containing the variable **frames_LL** could be saved as frames_LiveLoad.mat

DLf = dead load factor.

Note: the dead loads, superimposed dead loads, live loads and wind loads must be combined. **HR_DAD_1.1** allows the user to combine these loads by defining load factors which are then used to define appropriate load combinations.

3D 2 story example: using the ASCE 7-05 Standard load combination:

1.2D + 1.0L + 1.6W

in which D is the dead and superimposed dead loads, L is the live load while W is the wind load, the appropriate value is therefore 1.2.

SDLf = superimposed dead load factor.

3D-2 story example: see above under **DLf** .

LLf = live load factor.

3D-2 story example: see above under **DLf**.

WLf = wind load factor.

3D-2 story example: see above under **DLf**.

g = local response peak factor.

41

Note: the local response peak factor is used to calculate the peak internal forces occurring in the various members. Appropriate values of this parameter are between 3 and 4.

3D 2 story example: an appropriate value would be 3.5.

END PAGE_ FOUR INPUT

INPUT:

interstory_location = load the file containing the position of the column line where the interstory drift is desired.

> *File structure*: The file can be constructed in MATLAB. The variable in the mat file containing the position of the column line must be named **interstory_location**. The mat file containing the variable **interstory_location** can then be saved under any name. The variable **interstory_location** is a matrix, the first 2 rows of which contain the x and y coordinates of the column line with respect to the center of mass of each floor. The third and last column contains the height of the story. The first row contains information relating to the first floor, while the second row stores information related to the second floor and so forth. Successive column lines are appended as an extra N rows (where N is the number of floors).

> *3D 2-story example*: wanting to consider the column line formed by members 7-8 and 5-6 (Fig. A1) the variable **interstory_location** is:

$$\textbf{interstory\_location} = \begin{bmatrix} 3 & 3 & 4 \\ 3 & 3 & 4 \\ 3 & -3 & 4 \\ 3 & -3 & 4 \end{bmatrix}$$

The mat file containing the variable **interstory_location** could be saved as Interstory_Drift_Input.mat

Qss = choose whether to consider observed peaks or peaks estimated as indicated in Sect. III B, www.nist.gov/wind.

Note: the default is to consider the observed peaks of the inter-story drift time histories.

acceleration_location = load the file containing the positions of the points belonging to the top floor where the peak acceleration is desired.

File structure: The file can be constructed in MATLAB. The variable in the mat file containing the position of the points must be named **acceleration_location**. The mat file containing the variable **acceleration_location** can then be saved under any name. The variable **acceleration_location** is a matrix, the first 2 rows of which contain the x and y coordinates of the point with respect to the center of mass of the top floor. The first row contains information relating to the first point while the second row stores information related to the second point and so forth.

3D 2-story example: If one of the corner points is considered, for instance the point with the coordinates $x = 3$ and $y = 3$, the variable **acceleration_location** is:

$$\textbf{acceleration\_location} = \begin{bmatrix} 3 & 3 \end{bmatrix}$$

The mat file containing the variable **acceleration_location** could be saved as Acceleration_Input.mat

Qs = choose whether to consider observed peaks or peaks estimated as indicated in Sect. III B, www.nist.gov/wind.

Note: the default is to consider the observed peaks of the top floor acceleration time histories.

The following pertains to the simulated 999 extreme wind events provided for a large number of locations (mileposts) along the Gulf of Mexico and North Atlantic coast. This simulated data is publicly available at http://www.nist.gov/wind by following the links for extreme wind data sets.

Hmp = hurricane milepost

> *3D 2-story example*: considering the building located in Miami the milepost is 1450.

flHfile = folder location of the database of simulated hurricanes.

> *3D 2-story example*: after downloading the simulated hurricane database place it into any folder on your machine and point **flHfile** to this folder.

Vth = minimum wind speed under which the response is no longer desired

> *Note*: If a low minimum wind speed is considered this will lengthen the calculation of the responses with a specified MRI. However the higher the minimum wind speed is the greater the possibility of not considering a critical wind effect is.

> *3D 2-story example*: a reasonable minimum speed is 15 m/s

> *Note:* In general several types of wind need to be considered – see Sect. 6, item 4 in this report. Software for inclusion of up to three types of wind is being developed. See also Grigoriu (2008).

END PAGE_ FIVE INPUT

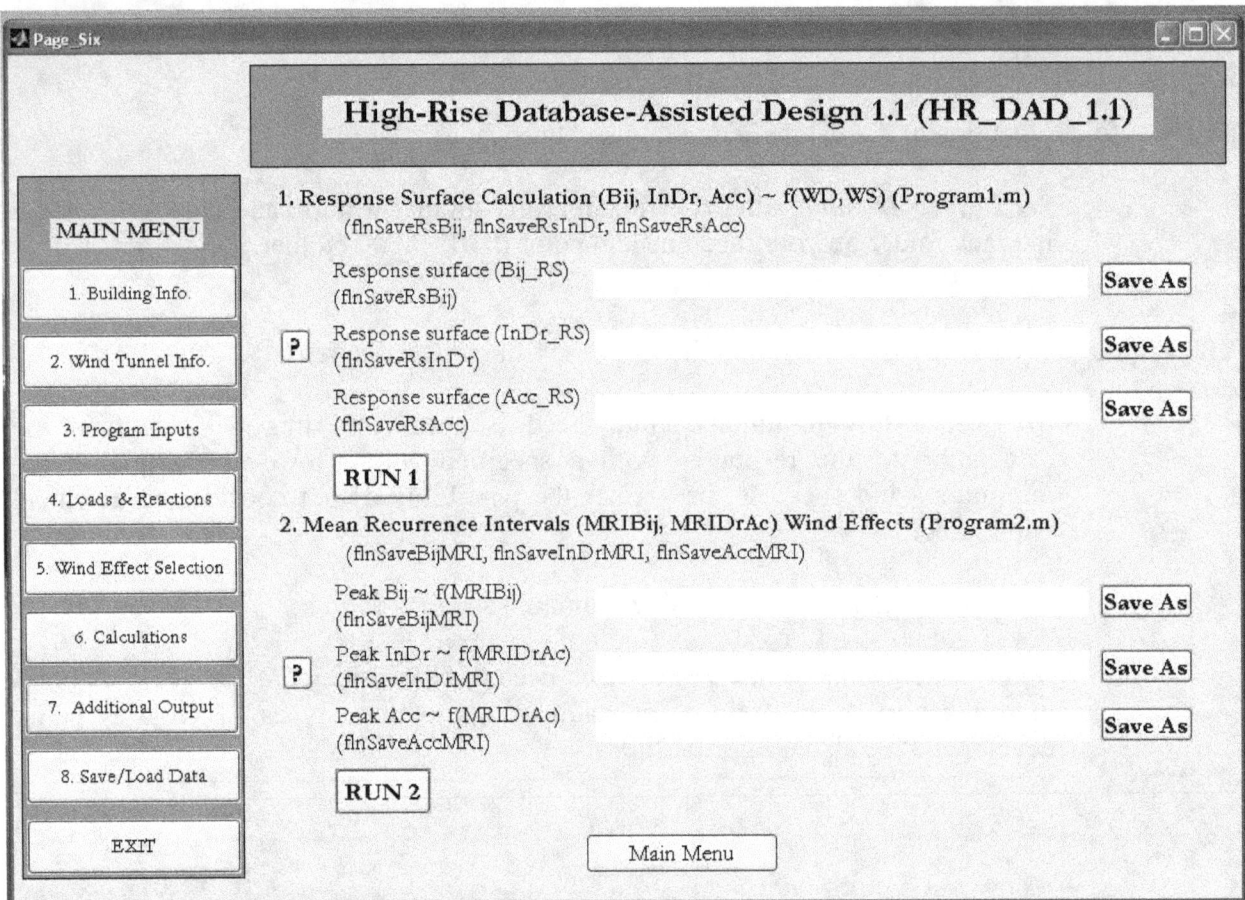

OUTPUT:

flnSaveRsBij = specify file location and name where the member response surfaces will be saved.

> *File structure*: The file that will be saved is a MATLAB mat file. The variable in the mat file containing the response surfaces will be named **Bij_RS**. The mat file containing the variable **Bij_RS** can then be saved under any name. The variable **Bij_RS** is a 3D array, each face of which represents the response surface of a specific member. The file structure is schematically represented in Fig. A4.
>
> In particular the index of each face corresponds to the column index of **mem_list** and therefore identifies the member. An element of a specific face identified by the indexes (*i,j*) corresponds to the peak demand/capacity ratio for the wind direction given by the *i*th element of **WD** and the wind speed given by the *j*th element of **WS**. Therefore each face will have a number of rows coinciding with the number of wind directions in **WD**, and a number of columns coinciding with the number of wind speeds in **WS**.

46

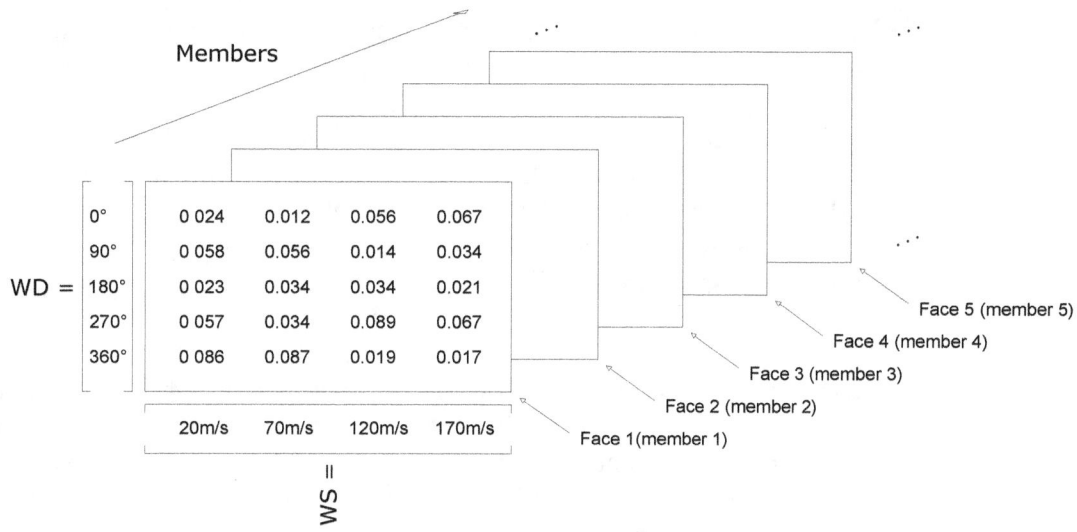

Figure A4. **Bij_RS** *file structure.*

3D 2-story example: plotting the first face of **Bij_RS** gives a graphical representation of the response surface associated with member 1. The plot is shown in Fig. A5.

Figure A5.Response surface associated with member 1.

flnSaveRsInDr = specify file location and name where the inter-story response surfaces will be saved.

File structure: The file that will be saved is a MATLAB mat file. The variable in the mat file containing the response surfaces will be named **InDr_RS_set_X** where X depends on the column number under consideration. The mat file containing the variable **InDr_RS_set_X** can then be saved under any name. The variable **InDr_RS_set_X** is a 3D array, each face of which represents the response surface in a specific direction (*x* or *y*). In particular the file is arranged so that the first *N* faces (*N* is the number of floors of the building) of the array are associated with the response in the *x*-direction while the next N faces are associated with the *y*-direction. The structure of the file is schematically shown in Fig. A6. An element of a specific face identified by the indexes (*i,j*) corresponds to the peak inter-story drift for the wind direction given by the *i*th element of **WD** and the wind speed given by the *j*th element of **WS**. Therefore each face will have a number of rows coinciding with the number of wind directions in **WD**, and a number of columns coinciding with the number of wind speeds in **WS**.

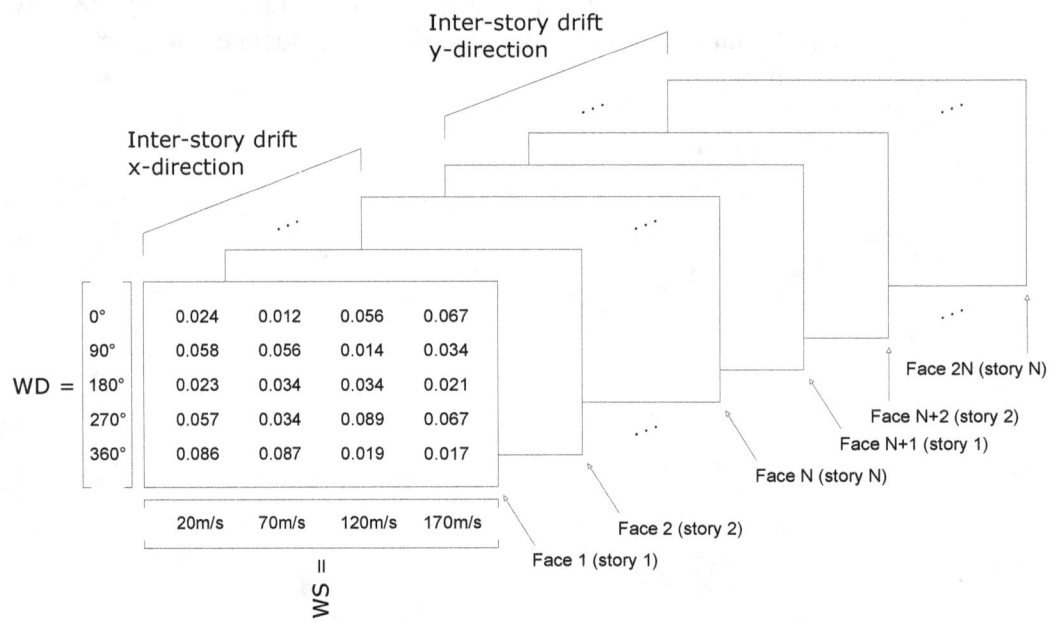

Figure A6. **InDr_RS_set_X** *file structure.*

3D 2-story example: plotting the second face of **InDr_RS_set_1** gives a graphical representation of the response surface associated with the inter-story drift in direction *x* of column line 1. The plot is shown in Fig. A7.

Figure A7. 2nd x-direction floor response surface of set 1

flnSaveRsAcc = specify file location and name where the top floor acceleration response surfaces will be saved.

File structure: The file that will be saved is a MATLAB mat file. The variable in the mat file containing the response surfaces will be named **Acc_RS_point_X** where X depends on the point belonging to the top floor. The mat file containing the variable **Acc_RS_point_X** can then be saved under any name. The variable **Acc_RS_point_X** is a 3D array, each face of which represents the response surface in a specific direction (*x* or *y*). In particular, the file is arranged so that the first face of the array is associated with the response in the *x*-direction while the next face is associated with the *y*-direction. The structure of the file is schematically shown in Fig. A8. As can be seen an element of a specific face identified by the indexes *(i,j)* corresponds to the peak inter-story drift for the wind direction given by the *i*th element of **WD** and the wind speed given by the *j*th element of **WS**. Therefore each face will have a number of rows coinciding with the number of wind directions in **WD**, and a number of columns coinciding with the number of wind speeds in **WS**.

49

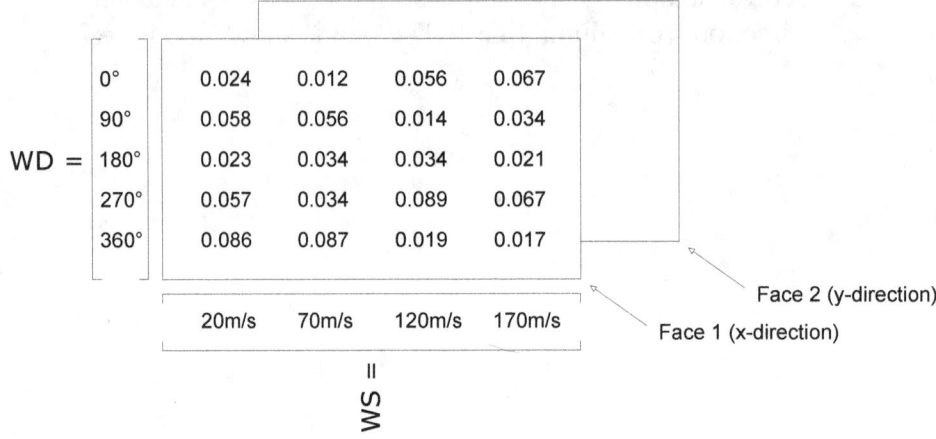

Figure A8. **Acc_RS_point_X** *file structure.*

3D 2-story example: Plotting the faces of **Acc_RS_point_1** gives results similar to those shown for the member response surfaces and inter-story drift.

flnSaveBijMRI = specify file location and name where the member responses with specified Mean Recurrence Intervals (MRIs) will be saved.

File structure: The file that will be saved is a MATLAB mat file. The variable in the mat file containing the member responses with specified MRIs will be named **Bij_MRIs**. The mat file containing the variable **Bij_MRIs** can then be saved under any name. The variable **Bij_MRIs** is a matrix the rows of which contain the peak capacity/demand ratios for a specific MRI. In particular the first row will contain the ratios corresponding to the first MRI contained in **MRIBij**, the second row those corresponding to the second MRI of **MRIBij** and so forth. The index of each column corresponds to the column index of **mem_list** and therefore identifies the member.

3D 2-story example: considering the MRIs input on page 3, **MRIBij** = [20 50], **Bij_MRIs** takes on the following form:

$$\mathbf{Bij\_MRIs} = \begin{bmatrix} 1.11 & 0.41 & 1.35 & 0.49 & 1.33 & 0.49 & 1.18 & 0.43 \\ 1.33 & 0.49 & 1.66 & 0.60 & 1.62 & 0.59 & 1.39 & 0.51 \end{bmatrix}$$

Therefore **Bij_MRIs(1,1)** = 1.11 means the for member 1 the maximum demand/capacity ratio with a MRI of 20 years is 1.11.

flnSaveInDrMRI = specify file location and name where the inter-story drift responses with specified Mean Recurrence Intervals (MRIs) will be saved.

File structure: The file that will be saved is a MATLAB mat file. The variable in the mat file containing the inter-story responses with specified MRIs will be named **InDr_MRIs_set_X** where X depends on the column number under consideration. The mat file containing the variable **InDr_MRIs_set_X** can then be saved under any name. The variable **InDr_MRIs_set_X** is a 3D array, each face of which represents the inter-story drift response, for column line X, with specified MRI. In particular the first face will contain the responses corresponding to the first MRI contained in **MRIDrAc**, the second face those corresponding to the second MRI of **MRIDrAc** and so forth. Each face has two columns corresponding to the *x* and *y* directions, respectively, and *N* rows which correspond to the *N* floors.

3D 2-story example: Considering the MRIs input on page 3, **MRIDrAc** = [10 20], the first face of **InDr_MRIs_set_1** takes on the following form:

$$\textbf{InDr\_MRIs\_set\_1}\ (face\ 1) = \begin{bmatrix} 0.025 & 0.026 \\ 0.015 & 0.016 \end{bmatrix}$$

Therefore **InDr_MRIs_set_1(1,1, face 1)** = 0.025 means that the peak inter-story drift in direction x of column line 1, floor 1 with MRI of 10 years is 0.025. In direction y floor 1, it is **InDr_MRIs_set_1(1,2, face 1)** = 0.026 and so forth.

flnSaveAccMRI = specify file location and name where the top floor acceleration responses with specified Mean Recurrence Intervals (MRIs) will be saved.

File structure: The file that will be saved is a MATLAB mat file. The variable in the mat file containing the top floor acceleration responses with specified MRIs will be named **Acc_MRIs_point_X** where X depends on the point belonging to the top floor. The mat file containing the variable **Acc_MRIs_point_X** can then be saved under any name. The variable **Acc_MRIs_point_X** is a 3D array, each face of which represents the top floor acceleration response, for point X, with specified MRI. In particular the first face will contain the responses corresponding to the first MRI contained in **MRIDrAc**, the second face those corresponding to the second MRI of **MRIDrAc** and so forth. Each face has two columns corresponding to the *x* and *y* directions.

3D 2-story example: Considering the MRIs input on page 3, **MRIDrAc** = [10 20], the first face of **Acc_MRIs_point_1** takes on the following form:

$$\textbf{Acc\_MRIs\_point\_1}\ (face\ 1) = \begin{bmatrix} 0.0319 & 0.0326 \end{bmatrix}$$

Therefore **Acc_MRIs_point_1 (1,1, face 1)** = 0.0319 means that the peak top floor acceleration in direction *x* of point 1 with MRI of 10 years is 0.0329, while in direction *y* it is **Acc_MRIs_point_1 (1,2, face 1)** = 0.0326 and so forth.

END PAGE_ SIX OUTPUT

PAGE_SEVEN *(Additional output)*

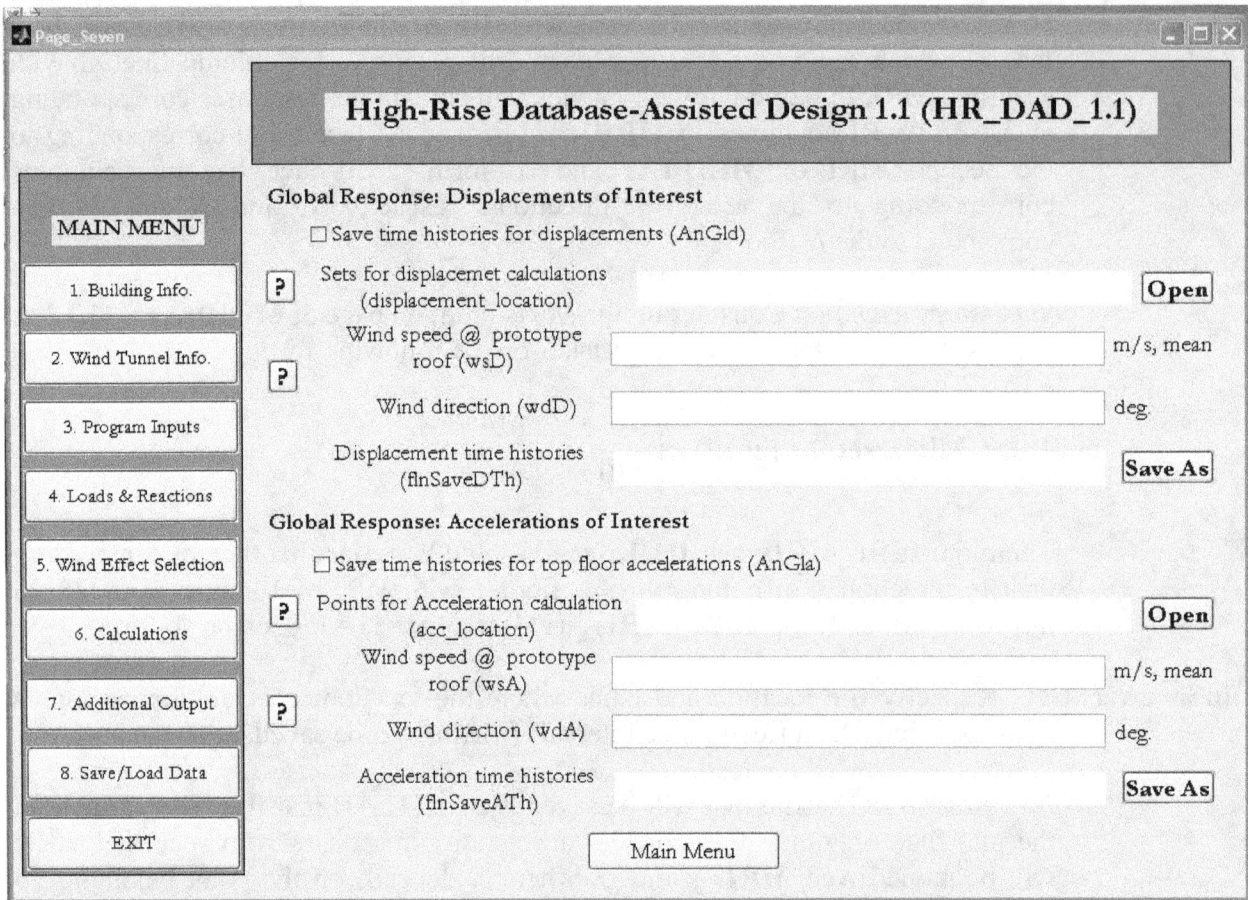

DISPLACEMENTS:

AnGld = choose whether to save select time histories of displacements.

displacement_location = load the file containing the position of the column line where the time histories of the displacements are desired.

> *File structure*: The file can be constructed in MATLAB. The variable in the mat file containing the position of the column line must be named **displacement_location**. The mat file containing the variable **displacement_location** can then be saved under any name. The variable **displacement_location** is a matrix, the first 2 columns of which contain the *x* and *y* coordinates of the column line with respect to the center of mass of each floor. The first row contains information relating to the first floor while the second row stores information related to the second floor and so forth. Successive column lines are appended as an extra *N* rows (where *N* is the number of floors).

> *3D 2-story example*: If the column line formed by members 7-8 (Fig. A1) is considered, the variable **displacement_location** is:

$$\text{displacement\_location} = \begin{bmatrix} 3 & 3 \\ 3 & 3 \end{bmatrix}$$

The mat file containing the variable **displacement_location** could be saved as Displacement_Input.mat

wsD = wind speeds at roof height of the full scale building for which the displacement time histories are desired.

> *3D 2-story example*: For example, if 30m/s, 40m/s are desired then the following vector would be entered:
>
> [30 40]
>
> *Note:* The variable must be input using square brackets as shown and must be *equal or a subset* of **WS**

wdD = wind directions for which the displacement time histories are desired.

> *3D 2-story example*: For example, if 40° and 180° are desired, then the following vector would be entered:
>
> [40 180]
>
> *Note 1:* The variable must be input using square brackets as shown and must be *equal to or a subset* of **WD**
>
> *Note 2:* The variables **wsD** and **wdD** *must* have the same lengths. Indeed the time histories of the displacements are calculated for pairs of wind speeds and directions.

flnSaveDTh = specify file location and name where the time histories of the displacements will be saved.

> *File structure*: The file that will be saved is a MATLAB mat file. The variable in the mat file containing the displacements time histories will be named **disp_set_X_wdD_wsD** where X depends on the column line while **wdD** and **wsD** will take the values associated with the time history. Therefore for each wind direction and speed a separate mat file will be generated. The mat file containing the variable **disp_set_X_wdD_wsD** can then be saved under any name but will end with **_set_X_wdD_wsD** indentifing the column line and wind speed and direction. The variable **disp_set_X_wdD_wsD** is a 3D array, each face of which represents the displacement time histories in a particular direction. The first face is associated with the *x*-direction, the second with the *y*-direction while the third and last is associated with the rotation around the *z*-axis. The faces are arranged so that

each row contains a time history. The first row is associated with the first floor, the second row with the second floor and so forth.

3D 2-story example: Having considered a single column line (members 1-2, Fig. A1) with two sets of wind speeds and directions, two mat files will be saved containing the variables **disp_set_1_40_30** and **disp_set_1_180_40**. If it is desired to display the time histories associated with the displacements in the *x*-direction at the top of column line 1, for a wind speed of 40m/s and direction of 30°, it is simply necessary to plot **disp_set_1_40_30(2,:,1);** this constitutes the second row of the first face. The time history is shown in Fig. A9.

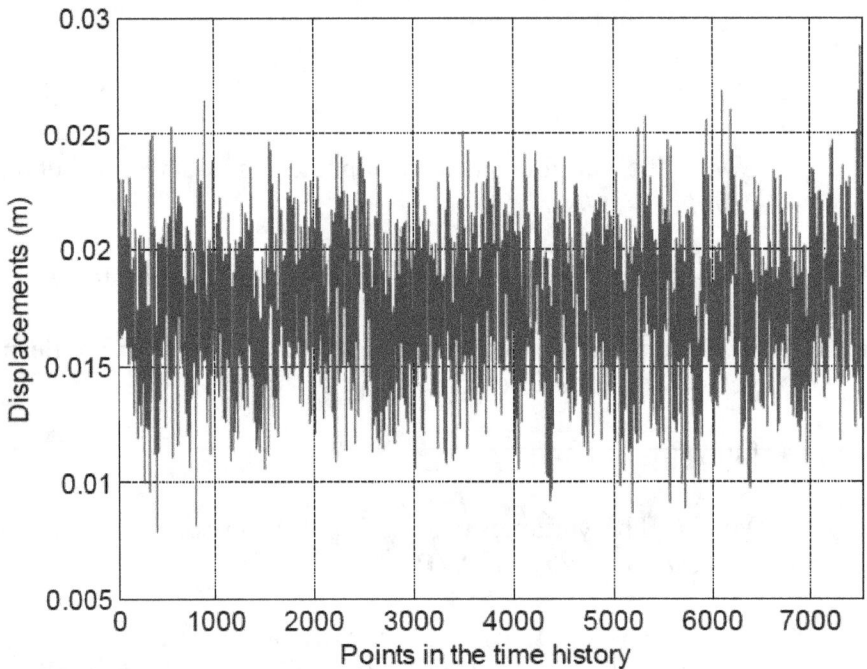

Figure A9. **disp_set_1_40_30(2,:,1)**, *2nd floor x-direction displacement time history for column line 1.*

TOP FLOOR ACCELERATION:

AnGla = choose whether to save select time histories of top floor accelerations.

acc_location = load the file containing the positions of the points where the time histories of the top floor accelerations are desired.

File structure: The file can be constructed in MATLAB. The variable in the mat file containing the position of the points must be named **acc_location**. The mat file containing the variable **acc_location** can then be saved under any name. The variable **acc_location** is a matrix, the first 2 columns of which contain the *x* and *y* coordinates of the points with respect to the center of mass of the top floor. The first row contains information relating to the first point while the second row stores information related to a possible second point and so forth.

3D 2-story example: If the point above member 8 (Fig. A1) is considered, the variable **acc_location** is:

$$\textbf{acc\_location} = \begin{bmatrix} 3 & 3 \end{bmatrix}$$

The mat file containing the variable **acc_location** could be saved as Acc_Input.mat

wsA = wind speeds at roof height of the full scale building for which the top floor acceleration time histories are desired.

3D 2-story example: For example, if 35m/s, 45m/s are desired then the following vector would be entered:

[35 45]

Note: The variable must be input using square brackets as shown and must be *equal to or a subset* of **WS**

wdA = wind directions for which the top floor acceleration time histories are desired.

3D 2-story example: For example, if 70° and 250° are desired, then the following vector would be entered:

[70 250]

Note 1: The variable must be input using square brackets as shown and must be *equal to or a subset* of **WD**

Note 2: The variables **wsA** and **wdA** *must* have the same lengths. Indeed the time histories of the displacements are calculated for pairs of wind speeds and directions.

flnSaveATh = specify file location and name where the time histories of the top floor accelerations will be saved.

File structure: The file that will be saved is a MATLAB mat file. The variable in the mat file containing the top floor acceleration time histories will be named **Top_floor_acc_wdD_wsD** where **wdD** and **wsD** will take the values associated with the particular time history. Therefore for each wind direction and speed a separate mat file will be generated. The mat file containing the variable **Top_floor_acc_wdD_wsD** can then be saved under any name but will end with **_wdD_wsD** indentifing the wind speed and direction. The variable

55

Top_floor_acc_wdD_wsD is a 3D array, each face of which represents the top floor acceleration time histories in a particular direction. The first face is associated with the *x*-direction, the second with the *y*-direction while the third and last is associated with the rotation around the *z*-axis. The faces are arranged so that each row contains a time history. The first row is associated with the first point input form **acc_location**, the second row with a possible second point, and so forth.

3D 2-story example: wanting to evaluate the time history of the top floor acceleration in direction x of the point input in **acc_location** for a wind direction 70° and speed 35m/s, it is simply necessary to evaluate **Top_floor_acc_70_35(1,:,1)**.

END PAGE_ SEVEN

56

PAGE _EIGHT (Additional output)

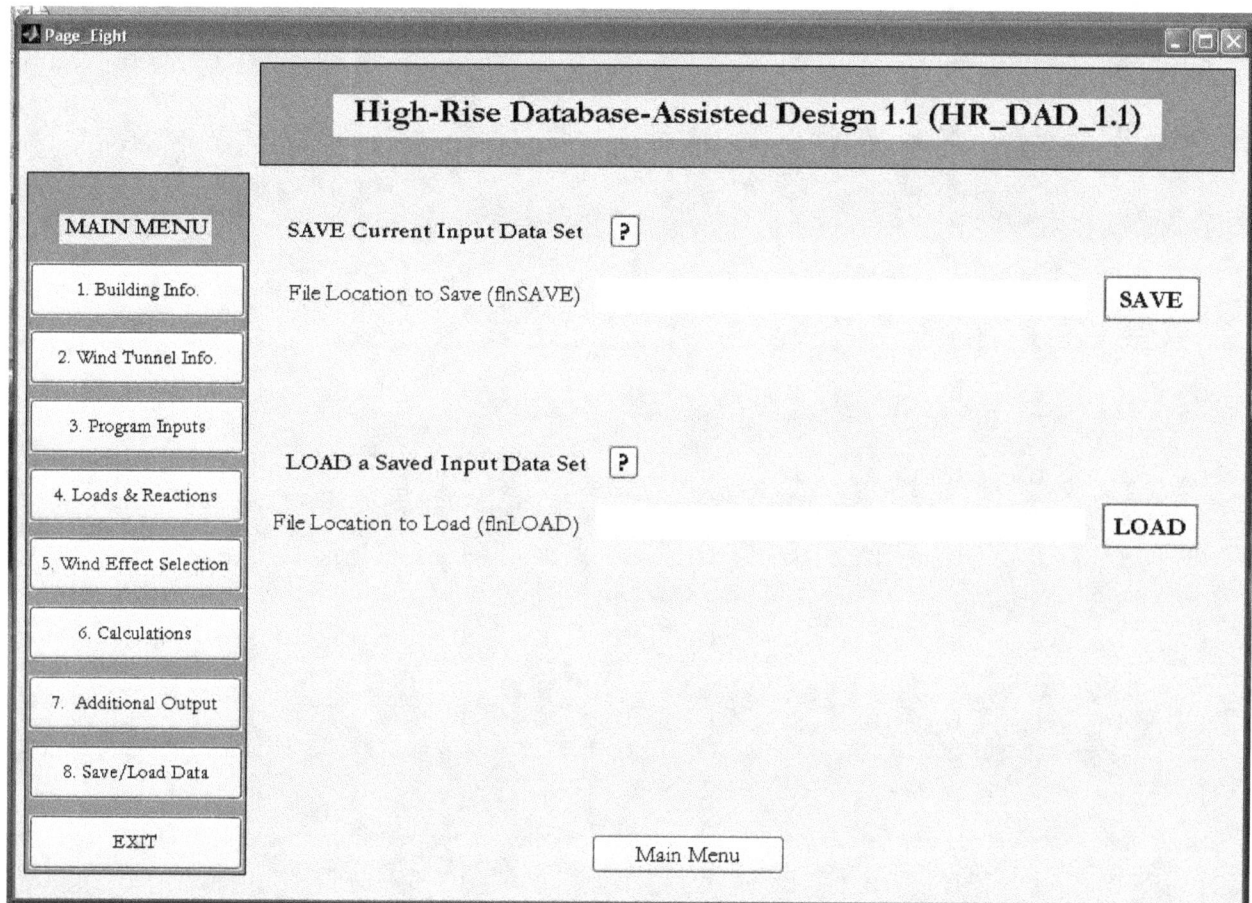

DISPLACEMENTS:

flnSAVE = specify file location and name where the all the input information loaded through
pages one to seven will be saved.

> *File structure*: The file is a MATLAB mat file that contains all the necessary
> information input at the time the file was saved

flnLOAD = load a file containing input information for running **HR_DAD_1.1**.

> *File structure*: The file is a MATLAB mat file generated during the definition of
> **flnSAVE**.

END PAGE_EIGHT